Contents

GW00420308

WRITTEN AND EDITED BY
BEN SMITH & KEVIN J. LEAR

SERIES CONCEPT & STYLE
KEVIN J. LEAR

DESIGNED BY BEN SMITH,
POCKET DESIGN STUDIO

PUBLISHED BY EXPLORING BRITAIN LTD,
THE GRANARY,
CARRYMOOR FARM,
HORNBY,
GREAT SMEATON,
NORTH YORKSHIRE,
DL6 2JF
TEL: (01325) 265015

Discover County Durham

Land of the Prince Bishops

County Durham is one of England's most northerly counties and part of the ancient border Kingdom of Northumbria. For hundreds of years the Prince Bishops of Durham ruled the County Palatine with a blend of ecclesiastical and political power unique in English history left behind a remarkable legacy which can still be traced in the 21st century.

Today, County Durham's fascinating blend of Christian, social and industrial heritage, combined with some of England's most beautiful and unspoilt landscape, will surprise and delight the most discerning visitor.

DURHAM CITY

Durham City is one of the most exciting visual and architectural experiences in Europe. For over nine centuries the magnificent Norman cathedral and castle have dominated the town from their dramatic location high above the River Wear. The castle and cathedral are now a World Heritage Site. The massive cathedral, "Half church of God, half castle 'gainst the Scot", is widely acclaimed as one of the world's finest buildings. It was built to house the shrine of St Cuthbert. Since then the cathedral and St Cuthbert's shrine have attracted pilgrims and travellers from all over the world.

Founded in 1072, Durham Castle remains one of England's largest and best preserved Norman strongholds and one of its grandest surviving Romanesque Palaces.

Downhill from the castle and cathedral you can still trace Durham's Mediaeval layout in its narrow winding streets and vennels. There is a wide range of shops, including the refurbished Victorian Indoor Market and the

VIEW FROM UPPER TEESDALE

Prince Bishops development. Away from the bustling town centre the shady wooded riverbanks offer a haven of peace with pleasant walks and boat trips on the River Wear.

The Oriental Museum is a treasure house of Oriental art and antiquities whilst the University Botanic Garden has trees and plants from all over the world. In contrast Crook Hall and Gardens are quintessentially English with a maze and Shakespeare Garden.

THE DURHAM DALES

Lying between the Yorkshire Dales and Northumberland National Park, the Durham Dales form part of the North Pennines, one of England's largest Areas of Outstanding Natural Beauty. The Dales, which make up over half of the County, offer some of the country's finest scenery - a blend of rugged upland, impressive waterfalls, gentle river valleys, wildflower meadows and drystone walls.

Teesdale, with its wooded valleys and characteristic white-washed farm buildings, has long inspired artists such as Cotman and Turner. At High Force the River Tees thunders 70 feet (21 metres) over massive rocks to form the largest waterfall in England. The Dales also boast some of the County's most attractive villages and towns. Historic Barnard Castle is full of character and home to one of County Durham's most remarkable attractions - The Bowes Museum. This splendid French-style chateau houses one of Britain's most important collections of European art. The bustling market town is named after its 12th century castle, now a craggy ruin overlooking the Tees. In contrast, nearby Raby still stands intact as one of England's finest Mediaeval castles. Both serve as reminders of this border area's turbulent past.

To the north is Weardale, once the hunting ground of Durham's Prince Bishops. Today quiet moorland roads open up panoramic Pennine views. In the 19th century the lonely hills of Upper Weardale buzzed with activity when the North Pennines was at the very heart of the world's lead mining industry. Today the lead industry has long since disappeared but the site at Killhope, The North of England Lead Mining Museum, now survives as a fascinating "hands-on" visitor experience.

Less well known than other upland regions, the area is surprisingly unspoilt and uncrowded, even at peak holiday times. There is plenty of scope for a whole range of outdoor activities amidst the impressive North Pennines scenery.

AROUND THE COUNTY

Around the County the visitor is constantly reminded of the area's rich social, industrial and Christian heritage.

Beamish, The North of England Open Air

KILLHOPE, THE NORTH OF ENGLAND LEAD MINING MUSEUM

Museum, is one of the North East's most popular and fascinating attractions. Staff in period costume invite visitors to experience the sights and sounds of the turn of the century Town, the Colliery Village with its Drift Mine, Home Farm and Pockerley Manor, all faithfully rebuilt from around the region.

As Beamish demonstrates, County Durham played an important role in the Industrial Revolution, especially in the development of the railways. Close to Beamish is Causey Arch, the earliest surviving railway bridge in the world, and the adjacent Tanfield Railway, the world's oldest existing railway. In 1825 the County saw the opening of the Stockton and Darlington Railway, the world's first public passenger line. At Shildon the home and workplace of Timothy Hackworth, the Railway's Superintendent Engineer, is now a Victorian and Railway Museum.

At Bishop Auckland is Auckland Castle, principal country seat of the Prince Bishops since Norman times. Just a few miles from Auckland Castle is Escomb Saxon Church. Dating from the 7th century it was built using stone from the nearby Binchester Roman Fort.

On the coast there are sandy beaches at Seaham and Crimdon and the Durham Coastal Footpath. Castle Eden, the largest of Durham's wooden coastal ravines, is rich in wildlife and now a National Nature Reserve. The County is also an excellent touring base, in close proximity to Hadrian's Wall, Holy Island (Lindisfarne) and the Northumberland Coast, Scottish Border Country, the Yorkshire Dales, the North York Moors and the Lake District.

For a free copy of the County Durham Holiday Guide contact: -

Tourism Team ,EDPD
County Hall, Durham City
County Durham, DH1 5UF
Telephone: (0191) 383 3354 (24 hours)
Fax: (0191) 383 3657
web Site: www.durham.gov.uk/tourism
e-mail: tourism@durham.gov.uk

DURHAM CASTLE
WORLD HERITAGE SITE

Palace Green, Durham, DHI 3RW Tel: 0191 374 3863
www.durhamcastle.com

The Castle is a superb venue for Wedding Receptions, Banquets, Dinner Dances, Conferences and Seminars and Holiday Accommodation which includes The Bishop's Suite with sitting room overlooking the River Wear, the walls of which are hung with 17th century tapestries. The bedroom has a four-poster bed and there is an adjoining bathroom.

In medieval times, the Prince Bishop's of Durham were temporal as well as spiritual lords, and they ruled and defended the north of England with vicegeral powers. The Bishop's seat was Durham Castle, which was founded by William the Conqueror in 1072. The Castle commands the narrow neck of the loop formed by the River Wear, which almost surrounds the steep wooded peninsula on which both the Castle and Cathedral stand. Since 1832 the Castle has been occupied by University College, the foundation College of the University of Durham.

The Castle with the Cathedral has the status of a World Heritage Site. This gives international recognition to the unique, historic and scenic character of this central part of Durham City. The status is only granted sparingly and put this site on a par with such world-famous buildings as the Taj Mahal.

Durham Castle is open to the public for guided tours as follows: July to September, Easter and Christmas vacations: daily from 10.00 to 12.00 noon and 2.00 - 4.30 pm; October to June inclusive, Monday, Wednesday, Saturday and Sunday afternoon from 2.00 - 4.30 pm. Special arrangements can be made for parties and specialist groups. Visitors should note that the Castle is closed during the Christmas period. University ceremonies and events also make it necessary to close the Castle at dates and times which cannot be specified in advance.

AUCKLAND CASTLE

THE SEAT OF THE BISHOP OF DURHAM

In mediaeval times the Bishops of Durham were indeed princes as well as bishops, invested with regal powers (bishops 'palatinate'), maintaining a string of castles and manors throughout the North as they exercised secular as well as ecclesiastical authority. Auckland Castle was then the principal country seat of the Prince Bishop, with Durham Castle his chief official residence. Bishop van Mildert was the last of such Prince Bishops: the royal authority of the Palatine was passed back to the Crown on his death in 1836. It was he who transferred that 'magnificent town house', the Castle adjacent to the

Cathedral at Durham, to the University of Durham which he founded in 1832. Hence Auckland Castle is now the only residence of the Bishop of Durham, and accommodates his office as well as the central administration of the Diocese. With (some currently existing) parts of Auckland Castle dating back over eight centuries, the building has had a complex history. It was in the 12th century that Bishop Hugh de le Puiset first made this fortified manor house the Bishop of Durham's principal country residence. Thereafter successive bishops have added buildings, or made their mark on it in other ways. The consequent variety of its architecture is immediately striking as one leaves the drive and passes through the screen, this was created in 1796 for Bishop Shute Barrington, whose coat of arms, with those of the Diocese, is to be seen on the central arch, James Wyatt, its architect, was also responsible for much of the interior alteration and decoration of the castle that we see today.

Believed to be the oldest part of the earlier fortified house, the Entrance Hall, once known as the Gentlemen's Hall, originally linked the Servant's Hall and Kitchen with the 12th century Great Banqueting Hall (now the Chapel). The ornate (and false) ceiling hides the original massive oak beams. Its elegance is echoed in the slender but imperial double-return staircase, branching to right and left.

The Throne Room, gracious and well-proportioned as it is, is also imposing. It was meant to impress. For immediately upon entering it, one is facing at its far end the Bishop's Throne set against an ornate plaster screen depicting the arms of the Diocese in mid-blue, gold and silver, impaled with those of the 18th century Bishop Barrington, supported by a crook and a sword, and surmounted by a bishop's mitre rising from a ducal coronet. Around the walls of this room hang portraits of some of the successive bishops who have occupied this See. In sheer scale the dominating portraits are probably those of Barrington and van Mildert, but from a much more recent - and very different - era there are impressive portraits of Bishop Michael Ramsey, Bishop David Jenkins and Bishop Michael Turnbull. The two hundred year-old tinted windows by James Wyatt are of very pale green and

pink glass - 'to make the ladies appear less pale in the bright sunlight'.

The door on the right of the Bishop's Throne leads into the next State Room, the Long Dining Room. This part of the Castle was added in the 16th century, some four hundred years after the original foundation. This Long Dining Room has much the same impressive proportions as the Throne Room it adjoins. Originally the work of Bishops Ruthall and Tunstall in the 16th century, it has been considerably modified by later bishops.

The King Charles Room is of the same period as the Long Dining Room, this room; originally built as a State Bedroom during Bishop Tunstall's episcopacy, it takes its name from its most famous visitor, Charles I, whose bedroom it was when he rested here on his journeys to and from Scotland. His last call at the Castle, so the sad tale goes, was on the 4th February 1647, as a prisoner on his way to London, his subsequent trial, and execution. But his guards were unable to gain access. So the King was forced to

spend the night in the nearby public house of one Christopher Dobson, in Silver Street by the Market Place.

The Music Room for many years was used as a Music Room, though it now serves as a dining room.

St. Peter's Chapel was originally built in the 12th century by Bishop Hugh de le Puiset as a great Banqueting Hall, complete with buttery, wine cellar and minstrel gallery. A century later Bishop Beck added buttresses to the outside walls and carried out other reconstruction work. For nearly five hundred years the Prince Bishops of Durham royally entertained their guests here.

In its eight hundred years Auckland Castle has been a meeting place for many kinds of gathering, ecclesiastical, political, military and social. Today it remains a meeting place; but now it is one that makes available for public use the traditional splendour of the original reception rooms as well as a newly developed Conference Suite with its own access and reception.

Contact: The Manager, Auckland Castle, Bishop Auckland, Co. Durham DLI4 7NR
Tel: 01388 601627 Fax: 01388 609323
e-mail: auckland.castle@zetnet.co.uk
website: www.auckland-castle.co.uk

Timothy Hackworth
Victorian & Railway Museum

Hackworth Close, Shildon, County Durham DL4 1PQ Tel/Fax: 01388 777999

County Durham is famous as being the birth place of the railways, thanks to the famous Stockton and Darlington railway. Timothy Hackworth was superintendent engineer from 1825 on the line and in 1833, realised the huge potential of rail travel and he developed his own engineering works. He resigned to develop the famous Soho Engine works at Shildon, and make his own locomotives. It was here that the first trains to run in Russia and Nova Scotia were built, and many ships were powered by marine engines designed and built on the premises. Now the whole 15 acre complex, plus his house, form the nucleus

for the Timothy Hackworth Victorian and Railway Museum, and proposed new National Railway Museum Out-Station which gives a fascinating insight into the early days of rail and steam power in England. Timothy was a true son of the North East, having been born at Wylam-on-Tyne in Northumberland in 1786, and dying in 1850 in the house that now forms part of the museum. He was a born engineer/blacksmith employed in a lot of early developmental work at Wylam including the building of 'Puffing Billy'. Whilst contracted at the Stephenson works in Newcastle he built 'Locomotion No.1' which later officially opened the Stockton and Darlington Railway Company. Thanks to him, Shildon became the first railway town in the world. Now it attracts thousands of tourists each year who want to find out about the transport revolution that took place in the early nineteenth century.

You don't have to be a railway buff to enjoy the Timothy Hackworth Museum, it has plenty of interactive activities for children and period room settings, working engine sheds and live steam train rides. The museum won the "Pride of Northumbria" award in 1999.

Opening Times: Wednesday to Sunday from Good Friday until last Sunday in October 10.00 - 17.00 daily; Admission Charge.
www.hackworthmuseum.co.uk

THE BOWES

——— MUSEUM ———

Barnard Castle, Co. Durham, DL12 8NP Tel: 01833 690606

The Bowes Museum - Bringing Art to Life

The first glimpse of the imposing Château of The Bowes Museum never fails to impress. The Museum, founded in 1892 by John and Josephine Bowes, houses one of Britain's finest collections of paintings, ceramics, antique furniture and textiles. Alongside the extensive permanent collection, the Museum hosts an active programme of both historical and contemporary exhibitions, featuring works by renowned artists as well as loans from regional and national galleries. A busy and varied programme of musical concerts also runs throughout the year.

Annual events include a Family Fun Day, National Archaeology Day celebrations, outdoor theatre by (both local and national production companies) in the summer as well as craft fairs throughout the year.

To complete your visit, the Museum has a licensed café, shop, park and gardens. It is situated on the outskirts of the historic market town of Barnard Castle, providing the perfect gateway to the outstanding natural beauty of Teesdale. The Museum is open daily from 11am until 5pm.

For further information about the events and exhibitions at the Museum, please contact the Museum reception on **01833 690606** or visit **www.bowesmuseum.org.uk.**

Durham Wildlife Trust

Registered Charity no.501038

Low Barns Nature Reserve

Rainton Meadows Nature Reserve

Rainton Meadows Nature Reserve is located adjacent to Rainton Bridge Industrial Estate close to the A690. The Reserve provides a wide variety of habitats including grassland, scrub, mature woodland and ponds. Facilities within the Reserve include disabled access, a wildlife garden and pond dipping area.

In the Visitor Centre there are toilets, a shop and café, which sells light refreshments. The exhibition area in the café provides information on Durham Wildlife Trust and local wildlife. There is also a purpose built classroom with educational resources.

Low Barns Nature Reserve

Low Barns Nature Reserve is near Witton-le-Wear village and borders on the River Wear. The site is signposted from the A689 between Crook and Bishop Auckland. It provides a wide variety of habitats including grassland, scrub, young and mature woodland, ponds, streams and a large lake with several islands.

Facilities within the Reserve include disabled access, picnic area, butterfly garden, bird hides, covered observatory, pond dipping platforms and a bird-feeding station in the winter.

The Centre provides toilets, a shop and café selling light refreshments. The

Bowlees Visitor Centre

exhibition area provides information on Durham Wildlife Trust and local wildlife. There is also a wormery and links into the Centre from cameras on the Reserve.

Bowlees Visitor Centre

Bowlees Visitor Centre is between Middleton-in-Teesdale and High Force waterfall, just off the B6277. Follow the road out of Middleton-in-Teesdale, past Newbiggin and the turning for the Centre will be on your right. The Visitor Centre is an ideal base for exploring Teesdale, as it is close to the River Tees, both High and Low Force waterfalls and Gibson's Cave.

Facilities adjacent to the Centre include public toilets, disabled access, picnic area and a trail leaflet, which takes you up to Gibson's Cave and is available from the Centre and in the Out and About section in this guide. The Centre provides a shop and café and an exhibition area on Durham Wildlife Trust and the Teesdale area.

Durham Wildlife Trust

Durham Wildlife Trust is a voluntary organisation which aims to conserve wildlife and promote nature conservation throughout County Durham, Sunderland, Gateshead, South Tyneside and Darlington and to encourage local people to get involved in looking after local wildlife.

For further details:
Telephone 0191 5843112
e-mail durhamwt@cix.co.uk
www.wildlifetrust.org.uk/durham

Rainton Meadows Nature Reserve

Durham Dales Centre

Castle Gardens
Stanhope
Weardale
County Durham   Tel: 01388 527650
DL13 2FJ   Fax: 01388 527461
e-mail: durham.dales@durham.gov.uk

The Durham Dales Centre was opened in 1991 and offers a range of facilities for local people and visitors. Situated in Stanhope, historic market town and ancient capital of Upper Weardale, the rest of the North Pennines Area of Outstanding Natural Beauty is easily accessible from here. The Centre is set within what were formerly the gardens of Stanhope Castle, which was built in 1798 for the Gateshead MP Cuthbert Rippon. There is ample car, coach and cycle parking and wheelchair access is available to most facilities including the toilets. The walled gardens were described in the late 19th century as having extensive glasshouses. A gazebo (small viewing pavillion) is included in the Dales Centre gardens and is approached by a magnificent yew hedged walk. The Dales garden is designed to reflect a typical cottage garden in the dales. Traditional cottage garden plants have been used and these inevitably attract associated wildlife such as birds and frogs, which can be seen congregating near the pond. Interpretation boards around the site explain elements of the garden, and a children's animal trail is available from the Tourist Information Centre.

The Tourist Information Centre offers an accommodation booking service, information about UK holidays, the region

and local places to visit, what's on and theatre booking. The gift shop has a wide selection of cards and gifts, picture postcards and maps. A new exhibition looks at interesting features or "treasures" in Stanhope, Weardale and the North Peninnes Area of Outstanding Natural Beauty.

You can enjoy the friendly atmosphere of the Tea Room that specialises in homemade food, light meals and beverages. A special feature of the Tea Room is a series of interesting murals, depicting local scenes and events. They have all been designed and executed by local artists.

Within the pleasant courtyard setting are some craft workshops to which visitors are very welcome.

The Centre has recently expanded, a new Meeting Room is available to hire with disabled access and business units are available to rent.

The Centre is open 7 days a week and further details are available from The Centre Manager.

Tel: 01388 527650/526393

South Tyneside - It's all go!

"SPIRIT OF SOUTH SHIELDS" - IRENE BROWN

With an array of attractions past and present, a wonderful coastline graced by beautiful beaches and spectacular cliffs and bays, family entertainment, fine leisure and recreational facilities, a compact shopping centre with open air market, quality accommodation and excellent transport links, South Tyneside is a great place to visit.

HERITAGE ATTRACTIONS

South Tyneside is steeped in history and heritage and there are numerous historical attractions on the doorstep.
Arbeia Roman Fort in South Shields presents an opportunity to step back to Roman times. The Fort is part of the World Heritage Site of Hadrian's Wall and the stunning reconstructions of the West Gate, the Barrack Block and the Commanding Officer's House provide a complete picture of life in Roman Britain. South Tyneside was home to medieval Europe's greatest scholar, the Venerable Bede who lived and worked in the monastery of St. Paul, Jarrow, 1300 years ago. Today, visitors to Bede's World in Jarrow can discover what life was like for Bede and his fellow monks and how much of his work is of continuing

importance to us today. The museum offers plenty of activities to interest the whole family and features a reconstruction of an Anglo-Saxon farm complete with rare breeds of animals and replica buildings.

South Tyneside has become famous as the birthplace of Britain's most popular author, Catherine Cookson and the area provided inspiration for many of her best selling novels. South Shields Museum and Gallery offers an opportunity to find out more about the life of Catherine Cookson and to discover the influence of Land, River and Sea on the area's fascinating history.

South Tyneside has a proud maritime heritage and it is the home of the Nation's first purpose designed lifeboat. One of the original lifeboats, Tyne, is preserved at the Wouldhave memorial adjacent to South Marine Park. Her crews saved 1,028 stricken mariners in six decades of service in the 19th century.

COASTLINE

South Shields has some of the finest beaches in the region and South Shields seafront is a beautiful place to enjoy a stroll whatever the time of year. Littlehaven offers excellent sailing, windsurfing and angling opportunities, whilst the golden stretch of Sandhaven, with its backdrop of dunes is especially popular and has won the Blue Flag Award 2002 in recognition of excellent facilities, amenities and water quality.

Wonderful scenic walks, quiet coves, golden sands, family picnic spots and mile upon mile of outstanding natural beauty, make South Tyneside's coastline a delight for young and old alike. The wide grassy

THE CUSTOMS HOUSE

"Landing Lights" - Martin Richman

the River Tyne at the heart of the cobbled Mill Dam has now been transformed into the Borough's focal point for the arts, with two theatres, a cinema, art gallery and fine restaurant. The Art on the Riverside programme in South Tyneside provides art enthusiasts with a visual feast. The Conversation Piece, created by acclaimed Spanish sculptor Juan Munoz is located at Littlehaven with its superb views of the harbour and features 22 bronze life size figures. The lovely and evocative Spirit of South Shields at Market Dock represents the Tyne's rich seafaring past and is the figurehead of its future. After dark another art form comes into its own. Landing Lights is a beacon of coloured lights towering above the ferry landing, a shining reminder of the role the river plays in

sweep of the Leas from Trow Rocks marks the beginning of the National Trust's two mile stretch of coastline, reaching to Lizard Point and taking in Souter Lighthouse, now a major visitor attraction. Visitors can climb to the lighthouse tower or take a guided tour of the complex, which includes a 19th century lightkeeper's cottage, shop and tearoom serving delicious home-made food. One of the most amazing rock formations in Britain dominates breathtaking Marsden Bay and Marsden Rock is renowned for its sea bird colonies, especially kittiwakes, cormorants and fulmars. At Marsden, if you are very quiet you may just hear the moans and groans of the Grotto ghost, John the Jibber.

CULTURAL ATTRACTIONS
The former Custom House overlooking

It's a laugh a minute

SOUTER LIGHTHOUSE

linking local communities. Further works are planned in the near future.

South Tyneside Council stages a number of events throughout the year most notably the annual summer Cookson Festival. This hugely popular free event features live music, children's entertainment, parades, street theatre, summer brass concerts, and Sunday Spectaculars featuring a selection of big name acts.

LEISURE ACTIVITIES

Fitness fanatics will be spoilt for choice in South Tyneside. Temple Park Centre boasts a whole host of facilities to keep all the family entertained. The exciting leisure pool incorporates a wave machine, 50m Aqua Blaster slide, learner pool and a diving tank. There are also squash courts, steam rooms, state of the art fitness facilities, and Pirate's Playland. The private members club Springs is located in the heart of the town and provides first class facilities including swimming, sauna, steam, a comprehensive fitness suite and programme of classes for all abilities as well as a range of pampering beauty treatments.

South Tyneside has an excellent network of picturesque cycling trails and walking routes around the Borough. During the summer months a cycle hire service operates from the seafront providing visitors with an opportunity to experience all of South Tyneside's coastal attractions whilst participating in an environmentally friendly and healthy activity. South Shields is also the home of the Great North Run Seaside Finish, when the Leas becomes the destination for 50,000 runners.

ST. PAUL'S CHURCH

King Street in South Shields is home to a number of major high street names and it is just a short window-shopping saunter to one of the North East's finest open air markets.

NIGHTLIFE

As the sun sets, the town centre lights up. The folk of South Tyneside know how to let their hair down and the trendy clubs and wine bars of downtown South Shields are a magnet to fun seekers from all over the area. For a more sedate evening and a taste of real ale the Mill Dam area is the place to be. For those with young children, there are a number of family friendly pubs on the seafront with spectacular views of the coastline. One thing is for certain, whatever your choice,

SEASIDE PLEASURES

Picturesque South Marine Park with its lake is a haven for swans and ducks and a relaxing spot for a picnic and a stroll and families find it hard to resist a ride on the miniature steam railway.

No visit to South Shields is complete without a trip to Ocean Beach Pleasure Park. Fun, laughter and thrills are in plentiful supply at this popular seafront funfair, with rides and amusements suitable for all ages.

PICTURESQUE VILLAGES

For quieter moments, a visit to one of South Tyneside's picturesque villages is recommended. Leafy lanes, cosy pubs and quaint shops offer a haven of tranquility.

SHOPPING

Shopping in South Tyneside is a pleasant and enjoyable experience. Pedestrianised

BEDE'S WORLD PRESERVING FARMING HERITAGE

THE LEAS - STUNNING COASTLINE

you will not only have fun but you will also find friendliness as the Geordies' reputation for kindness to strangers is well earned.

EATING OUT

South Tyneside offers an array of restaurants serving food from around the World including Indian, Chinese, Italian, English, Mediterranean and American. Ocean Road in South Shields is renowned throughout the region for its amazing array of fine Indian restaurants offering excellent value for money. No seaside town would be complete without fish and chips and there are numerous restaurants and takeaways dotted along the seafront and Ocean Road. For those with a sweet tooth an ice cream on a sunny summer day takes some beating.

TRANSPORT

South Tyneside is an ideal base as there are excellent transport links to all areas of the region. The A1(M) and A19 link up with a modern network of urban dual carriageways that give access to all parts of the Borough. The Metro service which serves Tyneside takes travellers through the heart of Jarrow, Hebburn and South Shields. The service through to Sunderland is now in operation so visitors can make the most of their stay. North Tyneside is also readily accessible thanks to the Tyne Tunnel and the ferry service.

South Shields Tourist Information Centre
South Shields Museum & Art Gallery
Ocean Road
South Shields
NE33 2HZ
0191 454 6612

BEDE'S WORLD

Where history was made

Travel back in time 1300 years and discover Bede's World, a place where history was made. Here the extraordinary life and times of the Venerable Bede, are celebrated through a stunning interactive museum, Anglo Saxon Farm and much more.

Age of Bede Exhibition

Come face to face with the Kings of Northumbria from a time when the region was at the height of its power. Marvel at genuine artefacts and listen to the amazing wisdom of Bede. Discover a world where the monastery was the centre of all learning and home to supremely skilled craftsmen. Find out why Bede himself made such an impact during his lifetime that he is still acclaimed 1300 years later.

Gwyre - The Anglo Saxon farm

A delight for adults and children alike, our farm uses rare animal breeds, timber buildings and traditional crops to bring ancient times to life. See and feed animals such as wild pigs, sheep, goats and ducks, bred to be close to those in Bede's time.

St Paul's Church and monastic site

Just a short walk across a tree-lined park is the enchanting and peaceful place where Bede lived and worked 1300 years ago. Parts of St Paul's Church date back to the seventh century. Here you can see Christian craftsmanship both ancient and modern and the ruins of the medieval monastery.

The herb garden

Enjoy the scents and hues of our mature herb garden, based on a traditional monastic plan. It is filled with a great variety of herbs that were used for both culinary and medicinal purposes.

Jarrow Hall café

Recently renovated, this beautiful Georgian building houses a modern café serving freshly made delicious meals and refreshments all day long.

The shop

Packed full of beautiful gifts made by local craftsman and books, videos and music to inspire you, the gift shop is well worth a visit.

Opening Times
April-October
Mon-Sat 10.00-5.30
Sunday 12.00-5.30
November-March
Mon-Sat 10.00-4.30
Sunday 12.00-4.30
Closed Good Friday. Call for details of Christmas opening.
Bede's World is located just 2 minutes from the South entrance of the Tyne Tunnel, follow the brown signs. Bede Metro station is a 15 minute walk away and buses from Jarrow Metro Station regularly pass the entrance.
Admission charges apply.
For more information contact:
Bede's World, Church Bank, Jarrow, Tyne and Wear NE32 3DY
Tel: 0191 489 2106 Fax: 0191 428 2361
Email: visitor.info@bedesworld.co.uk
Website: www.bedesworld.co.uk
Bede's World is a registered charity no 1009881

TAVISTOCK AT THE GROTTO

Coast Road, South Shields, Tyne & Wear NE34 7BS Tel: 0191 4556060

Tavistock at The Grotto

Tavistock at the Grotto is a delightful restaurant - cum - bar situated on the coastline of South Shields, Tyne & Wear. Without exception it must rate as the most interesting - and haunted - public house in Great Britain.

Tavistock at the Grotto - formerly known as The Tam O'Shanter, The Marsden Marine Grotto and The Marsden Grotto - is actually situated in a cave within the Magnesian limestone cliff face at Marsden Bay. The story of how it came to be there is contained within a true adventure story of 'Indiana Jones' proportions.

Circa 1843, Jack Bates, a lead miner from Allenheads, in Northumberland, relocated to Marsden Bay with his wife Jessie. Here he took up employment in a local quarry until his retirement. In 1782 Jack and Jessie, forced out of their rented cottage by an unscrupulous landlord, were unable to find alternative accommodation. They subsequently took up residence in a nearby cave facing the North Sea. Turning a disaster into a blessing, the industrious Jack converted the cave into a quaint dwelling, furnishing it with items fashioned from driftwood and ship wreckage.

Due to the large number of curious sight-seekers who journeyed to visit 'the odd couple', the entrepreneurial Jack eventually turned their home into a small cafe and made a small fortune. He even blasted a set of steps from the shore up to the top of the cliff face to make access easier, thus earning the epithet which has remained his ever since, 'Blaster Jack'.

Jack and Jessie lived in the cave for ten happy years before dying within months of each other.

In 1826, Peter Allan, the landlord of The

The Grotto circa 1870

Highlander public house in nearby Whitburn, moved into the derelict cave and enlarged it considerably by blasting holes in the cliff face. He converted the dwelling into a public house. His true motives have only recently come to light, for we now know that Allan and his family were searching for treasure.

It had long been rumoured that, in the 5th century, Roman soldiers from the nearby Arbeia fort had hidden a stash of coinage and valuables in the caves for safe - keeping. It is possible that Allan found what he was looking for, because he made several mysterious trips to London; specifically, on at least two occasions, to the British Museum. While the front of the cave complex operated as a public house, Allan, his brother John and his sons were excavating deeper and deeper into the cliff in an effort to strike pay-dirt.

The Allan family vacated the inn some twenty years after Peter's death in 1849. It was subsequently managed by a coterie of local dignitaries and politicians who were operating as a syndicate. During the succeeding decades the inn saw dire times and high times. The inn's doors shut - some thought for the last time - in June 2000, but this fascinating historical oddity was purchased again, privately, and is now one of the region's finest restaurants and public houses.

The ghost of Peter Allan still haunts the inn, as does that of his wife Lizzie. Other apparitions include that of a black and white cat, a smuggler and Margaret Allan, daughter of Peter and Lizzie.

A pewter tankard - dating from circa 1836

The Smugglers Tankard

- used to be placed on the bar every evening (full of beer) 'for the smuggler to drink'. By morning it would usually be empty. Sadly the tankard went missing when the pub was temporarily closed and has never been seen since.

Tourists will not be disappointed if they visit Tavistock at the Grotto.

Sunrise at The Grotto

SOUTH SHIELDS MARKET PLACE
THE PLACE TO VISIT

South Shields Market is the focal point for the town centre and attracts thousands of visitors and bargain hunters from near and far.

The long-established Market offers all the best to be found in an open-air traditional market place. It is one of the oldest and most popular markets in the North-East. The market is held in a traditional square with the Eighteenth-Century Old Town Hall at its heart. The historic church of St Hilda stands on the south side of the

Market Place and there has been a church on the site since 647 A.D.

The present church dates back to 1790. One of the finest open air markets in the North East, it has been recently modernised to include new paved area and a tree lined border.

Each week, the Market Place is alive with the buzz of bargain hunters and the banter of stallholders. Markets are held every Saturday and Monday and sell everything from plants, flowers, fruit and food to good quality clothing, toys, handbags, and art works.

It is also held each Wednesday from Easter to Christmas with locally grown fruits and vegetables, meat and dairy products, home made cakes, wooden furniture and

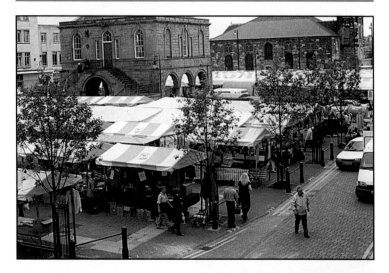

arts and crafts on sale.

The Friday flea market has its own special appeal to people who like to rummage as it has curios, books and bric a brac, all at low prices as well as second hand bargains. The Market Place is dominated by the elegant 'pepperpot' Old Town Hall which stands at its centre, offering shelter and a handy resting place.

South Shields Market is within easy reach of the South Shields Ferry terminal linking to North Shields and the Metro station is nearby. With good road access and convenient parking for motorists the Market is a great attraction for visitors and local shoppers. The Market Place makes an important contribution to the success of the lively pedestrianised town centre with a range of High Street stores together with some excellent pubs and restaurants of all kinds of eating opportunities.

Welcome to Derwentside

Discover Derwentside, an area steeped in proud history and heritage, host to some of England's most beautiful landscapes and an excellent base from which to visit some of the regions biggest tourist attractions.

Unspoilt Countryside and Stunning Landscapes

Derwentside is centrally located in the North East of England, with it's northern boundary tracing the River Derwent along the Durham/Northumberland border. In the west of the District lies the North Pennines Area of Outstanding Beauty, with the picturesque reservoirs and heather clad moors. The Derwent Valley is one of the least known valleys in the region, with walks along the banks of the River Derwent. The Derwent Reservoir offers a leisurely day out, with water sports, hiking and spectacular views over the water. A variety of parks and picnic areas such as Allensford Country Park are located within the District, ideal for visitors to relax and enjoy some of Derwentside's most beautiful scenery.

Industrial Heritage

This includes the world's earliest railways to coal mining and steel production. Tanfield Railway is an unforgettable day of steam train rides, with an opportunity to visit the oldest existing railway bridge in Britain - Causey Arch. Hownsgill Viaduct, a monument to Victorian Heritage, was constructed in 1857, to carry the Stanhope and Tyne Railways. The Terris Novalis Sculptures, located on the former Consett Steelworks site, symbolise the changes experienced by Derwentside since the closure of the Steelworks. Beamish, The North of England Open Air

Museum, voted Living Museum of the Year 2002, offers an insight into life in the North East in the early 1800's and 1900's. An English Heritage Property - the 18th century Derwentcote Steel Furnace is the oldest surviving example of a cementation furnace in the country.

Towns and Villages

Visitors have the opportunity to call into Consett and Stanley Town Centres, which were the focal points of Derwentside's iron, steel and coal mining industries for over 130 years, but more recently have developed facilities to cater for residents and visitors alike. There are numerous historic villages throughout the district including Shotley Bridge, a charming spa town, once home of the German Sword Makers, and paper making capital of the North. The village of Castleside, located along the A68 tourist route to Scotland is ideally situated as a base for exploring the District. Other picturesque villages include Lanchester and Ebchester where Derwentside's Roman connections can be explored.

Roman Remains

Discover Derwentside's Roman heritage along the Dere Street Trail, which has a number of forts built along its length. Within Derwentside these are the Longovicium Roman Fort at Lanchester and Vindomora Roman Fort at Ebchester. All Saints Parish Church, partly built using stone from, Lanchester Roman Fort and Includes a Roman and Romanesque Chancel Arch.

Other Places and Attractions to Visit

Hall Hill farm is a great day out, set in attractive countryside; visitors have the opportunity to see and touch animals at close quarters. Mister Twisters, the exciting indoor children's play and party venue, is host to the regions largest state of the art

soft climbing frame. Diggerland is a unique Adventure Park, which provides an outlet for the fascination that children have for mechanical diggers. Set in a tranquil landscape is the National Trust Property Gibside Estate once the home of the Queen Mother's family. The Glass Gallery boasts unique designs from stained glass to textiles, with the opportunity to visit the studio of glass artist Maralyn O'Keefe. The Grey Horse Pub is one of the oldest public houses in the region, dating back over 150 years, serves beer and real ale made on the premises in the Derwentrose Micro Brewery. The Reptile Centre in Burnopfield is a visitor and re-homing centre, which cares for unwanted and abandoned reptile species.

Activities

Derwentside is an ideal location to pursue a variety of activities. With the conversion of old railway lines re-opened as walk ways,

cycle routes and bridal ways, visitors can now explore Derwentside by foot, bicycle and horseback as part of their active holiday experience. For the more adventurous, the Sea-to-Sea (C2C) cycle route, with its sculpture trail runs through the District, offering cyclists one of the most challenging rides. Water sports, fishing, numerous golf courses and activity centres, offer something for everyone visiting Derwentside. Take a hot air balloon ride over the Derwent Valley and view the District at its best.

For further information on the wide variety of activities in Derwentside, please contact:

Project Development Officer for Tourism
Economic Development and Regeneration Unit
Derwentside District Council
Civic Centre
Medomsley Road
Consett
Co. Durham
DH8 5JA
Telephone: 01207 218237
Email - info@virtualtourismcentre.com
Web site - www.virtualtoursimcentre.com

The Glass & Art Gallery

194, Medomsley Road (opp. Civic Centre) Consett, Co. Durham DH8 5HX Tel: 01207 583353

In this age of mass production, what could be more inviting than individually crafted items in glass, wood and textiles or original paintings?

This is what a visit to the Glass and Art Gallery offers, the chance to browse at your leisure and admire the beautiful things uniquely created by North East artists. Each item makes a wonderful gift, the kind it's hard to give away!

The Gallery is owned by Ron and Maralyn O'Keefe. Maralyn is a well known glass artist and has her studio at the gallery.

Most days you will be able to watch the artists working on stained and kiln formed glass, an art going back centuries which is currently enjoying a revival. This is more than simply retail therapy, as you come through the door of the Gallery, leave the world outside and gaze at the superbly presented paintings, wood carvings, pottery, hand blown glass and textiles. It's a delightful and uplifting experience and one you're sure to want to repeat.

www.glassdesign.co.uk

Log on for the latest up-dates & special offers

Opening Times:
Monday - Saturday 9.00 - 5.00
Sunday   11.00 - 4.00
The Glass and Art Gallery
194, Medomsley Road (opp. Civic Centre)
Consett,
Co. Durham DH8 5HX
Tel: 01207 583353
Fax: 01207 500218

TANFIELD RAILWAY

The oldest existing railway in the world

The Tanfield Railway has now been around for more than twenty five years and here is a brief summary of what has been achieved and what can be achieved in the future.

The railway depends largely on enthusiastic volunteers, who have given their time and money to build a working historical railway. The enterprise began with a derelict locomotive shed and three miles of old railway formation. Track, locomotives, carriages, wagons etc. were bought, four bridges were built, stations erected, together with a large workshop, a signal box, lines were laid, engines and carriages rebuilt, staff trained and thousands of minor tasks undertaken to create today's working railway.

Few people realise just how far back our railway history goes. Many will think of the Stockton and Darlington Railway or the Wylam colliery railway as perhaps the earliest lines. To some it is staggering that the Tanfield line dates back to 1725, and that since then at least a part of it has been in continuous use.

Ghosts & Legends of Northumbria

The good, the bad and the gruesome of ghosts and legends in Northumbria. St Mary's churchyard in Morpeth is the final resting place for the suffragette Emily Wilding Davidson who threw herself in front of the Kings Horse on Derby Day at Epsom in 1913. Had this happened a century earlier Miss Davidson's body might have been stolen to be used for medical research in Edinburgh. Body snatchers had found this a convenient place on the A1 for stealing fresh bodies until 1830 when a watchtower was erected to guard them.

The Cauld Lad of Hylton, so called as his ghost was often seen shivering, appears to be a cheeky lad. There are many stories about how this young stable lad met his death at the hands of the temperamental Baron of Hylton. The story goes that he had not brought his horse out when ordered and the Baron had slain him with a pitching fork. He may have been caught napping which could be a result of many a night lying awake dreaming of the Baron's daughter for whom he allegedly was madly in love. The servants of the castle would often see the young lad, who liked nothing better than to mess up the tidy kitchen by throwing plates about. They soon realised that if they left the kitchen messy the ghost would tidy it up for them. Perhaps as a thank you the servants left a green cloak and hood out for the ghost who appeared at midnight. He uttered the final words "Here's a cloak, here's a hood. The Cauld Lad o' Hylton will do no more good" and with that he was not seen again.

Haltwhistle appears to be the scene for two legends about the Lord Blenkinsopp.

The first concerns Bryan Blenkinsopp who lived at Blenkinsopp Castle many centuries ago. The young Baron would often boast that he would marry a lady who owned a chest of gold so heavy that even ten of his strongest men would not be able to carry it. He met his bride to be abroad whilst fighting in the crusade and bought her back to England where she learnt of her husbands boasts. She became worried that her husband had only married her for her wealth and is reported to have hidden the treasure in the grounds. Bryan either upset by this lack of trust or, perhaps it was the humiliation, left his wife and castle never to return. The Lady regretted her actions and in spite of efforts to trace him died a sad and remorseful woman. She has been seen haunting the ruined grounds waiting to guide her husband to the treasure so that she can finally lie to rest.

The next incident regarding a Lord Blenkinsopp happened at Bellister Castle. The Lord was very suspicious of people and did not trust strangers. However, one night he gave shelter and food to a passing minstrel in return for entertainment. As the night progressed the Lord became very suspicious of the minstrel and sent his steward to check on the guest. The minstrel was nowhere to be found so the

Lord sent out a search party with hunting dogs to find him. The minstrels body was found on the banks of the Tyne savaged by the dogs, as it was too late to call them off. Bellister is said to be haunted by the ghost of a man in grey with a bloody gash on his pallid face.

Haughton Castle is said to be haunted by the Ghost of Archie Armstrong a notorious clan chief who was imprisoned there during the reign of Henry VIII. Thomas Swinburne the Lord of Haughton Castle had forgotten to leave instructions for the prisoner to be fed and watered. He suddenly remembered when he found the keys to his cell in his pocket and galloped home to feed him. It was too late by the time he got back in desperation Archie is reported to have gnawed at the flesh of his own arm. For many years Archies' ghost haunted the castle, until a vicar, using a leather bible exorcised it. His ghost returned briefly when the bible was taken

to London to be rebound and has rarely been seen since it returned.

In the churchyard of St Cuthbert's at Bellingham is a long stone which marks a grave, the occupant of which is now a piece of local folklore. The story centres around Lee Hall on the banks of the North Tyne to the south of Bellingham. The hall was home to the Ridley family who, when facing the prospect of a hard Northumberland winter, left their country residence for a milder climate in London. In the winter of 1723 the house was left in the care of three servants, they were under strict instructions not to allow any guest or lodger into the house. One bleak and miserable afternoon, a pedlar called at the hall carrying with him an unusually long package and asked if he could have shelter for the night. True to their employers instructions the servants refused the pedlar, but allowed him to leave his heavy burden while he sought shelter elsewhere. As the

night grew dark one of the servants became increasingly suspicious of the pedlar's long pack which had been left in the kitchen of the house. While lighting a candle the maid swore she saw the package move.

She quickly alerted the other two servants. The older man laughed at her suspicion, but the younger man, not wishing to take any chances fetched his gun and shot at the lang pack. To his astonishment a cry was heard and blood began to ooze from the mysterious package.

When the Lang pack was opened, the body of a dead man was found inside wearing a silver whistle around his neck. It seems that the man had been brought to the hall as part of a plot. The plan was obvious, this man was going to break free from his package and open the door for fellow accomplices to burgle the household!

Durham seems to have more than its fair share of ghosts known as 'Grey Ladies'.

The Grey Lady of Durham Castle is said to be the wife of one of the former Prince Bishops of Durham. She haunts the black staircase, built by Bishop Cosin in 1662, tragically she fell down it to her death. She can be seen walking the staircase, although since she died the level of the staircase has changed, as it was originally only attached to the outside walls of the Castle, and was considered to be far too dangerous. The Grey Lady however seems blissfully unaware of the changes and still walks along at the stairs original level!

Another story of a lady with simple tastes in her choice of clothes concerns The Grey Lady of Crossgate. In 1346, the Battle of Neville's Cross was fought between English and Scottish armies. Among the men that fought and lost their lives on the English side, one left behind him a wife and new born baby. But his wife didn't give him her 'farewell' as she hadn't wanted him to enlist.

People driving their coaches and wagons up Crossgate Peth would often stop for a drink somewhere along the way... and notice a drop in temperature. As they continued their journey they would notice the presence of the Grey Lady, with her new born child, hitching a lift, staying sombre, sad and silent, until they reached Neville's cross where she would disappear, perhaps looking for the body of her husband on the old battlefield. She isn't regularly sighted now, however, since the demise of horse drawn vehicles. Perhaps she doesn't understand that cars could take her too, or maybe she found the body of her love after all.

More recently in the Crossgate area, a ghost of a young woman has been sighted, this time without a baby. This is said to be the ghost of a Victorian girl from a workhouse near Allergate who was murdered and then thrown down a flight of steps. Her attacker was a soldier, who later confessed to his crime years after the event while living abroad.

Now there are those who have some unkind things to say about musicians and their abilities, and often pipers are more maligned than others - you either love the sound of

the pipes, or you hate them - well beware of the Ghostly Piper Jimmy Allen who died in1810. He was a very talented player of the Northumbrian Pipes, he was even official piper to the Duchess of Northumberland for two years. But he liked to drink, and gamble, and he also liked pretty women, conning them out of money to support his drinking habit. As well as this he was a horse rustler, consequently he was almost continually on the run from various authorities. He travelled as far afield as India, the Dutch East Indies and the Baltic, making money from playing his pipes with his extraordinary skill.

The authorities caught up with him eventually, however, and he was caught in Jedburgh, Scotland, for stealing a horse in Gateshead. He was sentenced to death in 1803 and spent the last seven years of his life in a cell beneath Elvet Bridge in Durham City, his sentence having been reduced to life imprisonment.

He died a few days before a pardon came through from the Prince Regent. It is said that if you listen carefully, you may still hear his ghost playing the Northumbrian Pipes from the cell beneath the bridge.

Good Food & Real Ale

All too often people pass through the counties of Durham and Northumberland, whether it be on the East Coast main line or the A1 trunk road, gaze out of the window at the countryside and then pass on. It is well worth pausing to enjoy the rugged grandeur of its scenery, and although it appears to be a very sparsely populated place there are some great pubs and restaurants just waiting to be discovered. No matter what your tastes, traditional or exotic, you will find something that will satisfy your palette. Many of Northumbria's local dishes are meatless, a legacy from harder times when only the affluent could afford to buy meat. Whitley Goose for example, is a concoction of onions, cheese and cream, but no goose! Even the names of the local dishes have a distinctive ring to them such

as Pan Hacklety (a mouth watering combination of cheese and vegetables fried together) and, as the north-east has a reputation for growing fine vegetables, especially leeks, it is not surprising to find Leek Pudding-in-a-Cloot (cloth) on some menus. The local markets are an excellent place to visit for examples of local foods and delicacies and, although Pease Pudding is known throughout the country, it's true home is the north-east. A combination of boiled ham shank, onions, split peas and seasoning it is a great accompaniment to roasts, salads and adds extra interest to a sandwich!

With such an abundance of meat, poultry, fish and vegetables within the area, it's no surprise that you can eat well here.

No food is complete without the appropriate beverage to help it on its way! Whilst Northumberland is not the ideal place for vineyards, it does have a great selection of breweries that pride themselves on producing 'real ale' in the traditional way such as the Border Brewery at Berwick, the Northumberland Brewery at Bedlington and the Hexhamshire Brewery at Hexham. The Border Brewery opened in 1992 and has enjoyed tremendous success, and their Christmas beer - Rudolph's Ruin makes an annual appearance.

The Hexhamshire Brewery also dates from 1992 and is run by Geoff Brooker of the Dipton Mill Inn. Although this is indeed a 'micro-brewery' (producing beer for consumption in its own pub) it's a five-barrel plant that produces three regular beers - Shire Bitter, the best seller locally, Devil's Water and intriguingly named Whap Weasel! A Christmas special, Old Humbug also makes a welcome return every year.

Best bets for Hexhamshire beer on its home patch are the Feathers Inn at Hedley on the Hill, the Mile Castle on the Military Road near Haltwhistle and Dyvells Hotel (page 86) at Corbridge.

The Northumberland Brewery dates from 1996 when it was established at Ashington - it has since moved to Bomarsund, near Bedlington. The brewery makes a range of four regular beers - best-seller is the session strength Castles Bitter which is accompanied by Northumberland County, Secret Kingdom, Northumberland Best, and Bomar Bitter, which is named after Bomarsund, a Finnish island, which was the site of a Viking battle. The brewery also makes a special beer - Old Black Diamond Bitter for the Black Diamond pub at Ashington.

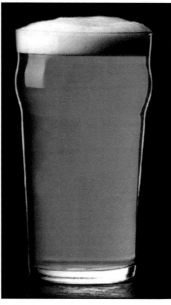

CARLBURY ARMS

PIERCEBRIDGE, DARLINGTON, CO. DURHAM DL2 3SJ - TEL: 01325 374286

The Carlbury Arms is an outstanding public house and restaurant found just off the main A67, five miles west of Darlington. The premises date back to the 17th century, as reflected in the original stone floors, exposed beams, sandstone walls and other features that have been retained throughout. Once known as The Wheatsheaf, it was given the name of the Carlbury Arms by previous owners who named it after a village that was once located here and of which only one house now remains, although the village is clearly depicted on a map of the area dated 1859. The immaculate white and blue building is deceptive, it looks small on the outside, but once through the door, you will be completely amazed at the feeling of space and comfort. Completely refurbished and redesigned, the interior of this charming pub is beautiful and welcoming. This is the epitome of the 'English Country Pub', whether you are a regular or a first-time visitor, the atmosphere draws you in. Open all day every day, food is served at lunchtime and evenings with the cosy and intimate restaurant (non smoking) seating 20. Meals are also served in the main bar area, however to avoid disappointment booking is advised for Wednesday - Saturday evenings and Sunday lunch. The menu includes traditional favourites such as home-made steak and kidney pie, gammon, chicken dishes, fresh fish and vegetarian choices. Depending on the season, there is also a choice of daily lunchtime and evening specials and wickedly tempting desserts. The bar is well stocked with a wide selection of traditional beers, real ales and an excellent wine cellar to complete your meal.

THE DOG INN

CROSS LANE ENDS HEIGHINGTON, DARLINGTON, DL2 2TX
TEL: 01325 312152 FAX: 01325 315689

The Dog Inn can be found just five miles from Darlington travelling north along the A 68. Enjoying a prime corner location with extensive well-kept grounds, The Dog Inn is easy to find for visitors who are unfamiliar to the area.

The charming stone building dates from the late 19th century when it was built as a hostelry to serve the local community. The Dog has a very inviting appeal and as visitors walk through the door for the first time they will quickly become aware of how friendly and welcoming an establishment it is.

The interior is well designed ensuring that you can always find a quiet corner for an intimate drink while also allowing plenty of room for smaller groups to gather. The decor retains many of the original features of the building with well-worn wood floors and low beamed ceilings. The restaurant has the original windowpane from the bar that was signed in 1916 by a local tradesman. Even the accommodation rooms were once the old blacksmith's shop, a picture of the last blacksmith tenant still hangs in the bar today.

Here you can enjoy some delicious food that has been prepared in the pub's kitchen. The menu offers an excellent choice of traditional dishes catering to various tastes and appetites. The bar stocks a wide range of beers and lagers with a choice of two cask ales.

If you are in need of accommodation you need look no further, as The Dog is also able to provide bed and breakfast. There is one single and two twin rooms, all with en-suite bathroom, colour TV and hot drinks facility.

The Three Tun's Inn

Eggleston - Barnard Castle - Co. Durham DL12 0AH
Tel: 01833 650289 e-mail: three-tuns-eggleston@tiscali.co.uk
www.three-tuns-eggleston.netfirms.com

Nestled in the beautiful dales lies
Eggleston, a picturesque and friendly
little village, beside the River Tees.
Overlooking the outstanding scenery is
The Three Tuns Inn.
Proprietors, Denis and Val Holmes offer
a warm welcome, and are always happy
to see new faces.
A traditional country pub, with wooden
beams, open fires, real ales and good
pub food, The Three Tuns is a place that
visitors return to again and again.
The light and airy dining room seats 60,
and has extensive views over Upper
Teessdale.

In the summer sit in our beer garden, relax and enjoy these views. In the winter, toast your toes in front of the log fires. The games room is well provided for, with pool, darts and a wide screen T.V.

Denis prides himself on his good old-fashioned home cooking, and customers can choose from what must be the widest menu selection in the dale, from the ever-changing specials board, to the extensive regular menu and light bites menu.

There's something special for everyone at The Three Tuns, Eggleston.

CROSS KEYS

**Thropton - Nr. Rothbury - Northumberland - NE65 7HX
Tel: 01669 620362**

Situated a couple miles from Rothbury, travelling towards Otterburn on the B6341 is the pretty village of Thropton, a journey well worth making for here you will find The Cross Keys Inn.

The area has some beautiful walks and magnificent scenery. This fine country hostelry dates back to the early 19th century and is hard to see clearly as it is covered with a sprawling red creeper. The building enjoys an elevated position, slightly higher than the rest of the village, and has some fine views.

The inn has a warm, welcoming atmosphere and serves a wide variety of fine ales and food of a very high standard which are regularly enjoyed by both the locals and visitors to the area. The cosy interior is kept warm and inviting on the cooler months by the open fires. Food is served each lunch time and evening from a varied menu catering to all tastes and at very reasonable prices. Recently taken over by a delightful couple, George Yule and Gale Hooper, they have brought new life and some fresh ideas to the place. George is a trained chef and Gale has lots of experience in working behind a bar, so there could well be some changes in years to come.

Should you wish to stay in the area longer, bed and breakfast accommodation is available throughout the year. There are three bedrooms furnished to a high standard, with en-suite, TV and tea and coffee making facilities at reasonable prices.

Historic Northumbria

Palace Green, Durham Cathedral - County Durham Economic Development Partnership

Not all houses that are of historic interest have to be grand manors or castles. There are many people who rose to fame, fortune and world acclaim from very humble beginnings. The North East has been the cradle of invention and innovation for centuries, and has seen the rise and success of many native sons and daughters.

Durham Cathedral.

When the Vikings attacked Holy Island, St Cuthbert's remains were removed by the monks and after much travelling were laid to rest in a loop of the River Wear - the present site of Durham Cathedral. St Cuthbert's coffin may still be viewed at the Cathedral which has been voted the most beautiful building in the world by an "Illustrated London News" survey of architects. Also at Durham Cathedral is St Cuthbert's Cross, a representation of which is used for the symbol of the St Cuthbert's Way long distance footpath between Melrose and Holy Island.

Wylam

George Stephenson's Birthplace. A small stone cottage, built around 1750, Stephenson was born here in 1781 and lived with his whole family in one room. He grew up amidst mining and iron foundries, surrounded by the materials that influenced his career as a brilliant engineer. Railway transportation was born on 27 September 1825 when Stephenson's "Locomotion" ran from Darlington to Stockton, carrying 450 persons at 15 miles per hour. Stephenson's later "Rocket" won a famous competition to find the fastest locomotive by travelling at an average speed of 36 miles per hour from Liverpool to Manchester in 1830.

Wylam was also the home of Charles Parsons, inventor of the multi-stage steam turbine, who lived at Holeyn Hall.

Lady Waterford Hall

Ford Village, off the A697, north of Wooler. A 19th century former village school beautifully decorated with biblical murals by Louisa, Marchioness of Waterford. All the faces were modelled by local villagers from the Ford and Etal estate.

Cherryburn, Mickley, Nr Stocksfield. Thomas Bewick, Northumberland's greatest artist and wood engraver was born in here in 1753. Within the cottage, there is an exhibition of his life and works in the same farmyard setting in which he

'Locomotion' replica - County Durham Economic Development Partnership

was raised. Bewick was a keen observer of all things natural, his love of wildlife was superbly translated into his etchings and engravings. On demonstration days, visitors can see the 18th century printing press at work - still using some of Bewick's original woodblocks.

Country" and there is an interesting South Shields Museum with a reconstruction of William Black Street where the author grew up. In 1929 she left the North East and established an apartment hotel in Hastings on England's south coast. She stayed in Hastings for almost 50 years, only returning to the North East of England in 1975. She lived in Northumberland, firstly at Corbridge and later at Langley near Haydon Bridge for 16 years until 1991 when she moved to Newcastle for her final years. She first started writing, in Hastings, to help herself overcome depression following several miscarriages.

Kirkharle Courtyard.

Birthplace of Lancelot "Capability" Brown who became the foremost English master of landscape gardening, whose works were characterized by their natural, unplanned appearance. He was thus responsible for destroying many splendid formal gardens which he replaced with open grasslands, lakes and informally placed trees. Situated a quarter of a mile off the A696 Newcastle to Jedburgh road at Kirkharle near Wallington

Catherine Cookson.

The romantic novelist, Catherine Cookson, is one of the best selling authors in the world. She was born and raised on the south banks of the Tyne at South Shields in an area now called South Tyneside. Today, the area often promotes itself as "Catherine Cookson

Cragside, Rothbury.

A Victorian house situated in 900 acre grounds which may be visited separately. Built by industrialist Lord Armstrong who became one of the great Victorian industrialists and engineers. He invented high pressure hydraulic machinery (including cranes) and revolutionised the design and manufacture of guns. His house was mainly designed by Norman Shaw and was the first in the world to be lit by hydro-electricity. It is now a showcase of

St Cuthbert's Window Durham Cathedral

Victorian art (especially Pre-Raphaelite paintings and wall coverings), architecture and technology.

Wallington, Near Cambo village.
A 17th century mansion, the much loved home of the Trevelyan family, with beautifully furnished interiors including a fine collection of dolls houses. The central hall is decorated with a stunning series of murals by William Bell Scott, depicting the history of Northumberland. The Hall also has strong associations with the poet Algernon Charles Swinburne (1837-1909) and is set in magnificent parklands and gardens which may be visited separately.

Brinkburn Priory, Weldon Bridge. between Morpeth and Rothbury. A deserted Augustinian Priory founded in 1135 and restored in the 19th century in a lovely setting beside the River Coquet.

Today it is one of the finest Gothic priory buildings in England. It is becoming an increasingly popular location for choral and organ recitals because of its superb acoustics, including an annual classical music festival

Hexham Abbey.
Founded by St Wilfrid in 674, the original Saxon church was (as many buildings of that time) constructed from stones taken from Hadrian's Wall. The crypt is all that remains of the original building and Roman inscriptions can still be seen on some of the stones. The present 12th century Abbey remains Hexham's Parish church. Its treasures include the stone "Saint Wilfrid's Chair", reputedly the coronation seat for the Kings of Northumbria; and the grand Midnight Stair which once led to the canon's dormitory.

Lindisfarne Castle, Holy Island.
Built on the orders of Henry VIII in 1550 (using stones taken from Lindisfarne Priory) to protect Holy Island from attack by the Scots, the castle was converted into a private home by architect Sir Edwin Lutyens in 1903. The rooms are filled with a fine collection of antique furniture, mostly oak, of the early 17th century. Ornaments and pictures are in harmony creating a unique and picturesque home. The castle's small walled garden was designed by Gertrude Jekyll and is only open when the gardener is in attendance.

Chipchase Castle.
Located on a minor road, 1.5 miles south east of Wark village in the North Tyne valley. An impressive medieval / Jacobean castle and pele tower set in 3 acres of formal and informal grounds with a specialist plant nursery.

Norham Castle, Norham village. 7 miles south west of Berwick.
The remains of one of the strongest of all border castles guarding a bridge across the River Tweed, with an outstanding Norman keep. Built by Bishop Pudsey in the 12th century, Norham was the northern outpost of the Prince Bishops of Durham who were given control over this part of Northumberland, known as Norhamshire. It is described by Sir Walter Scott in his poem "Marmion" as "the most dangerous place in Britain". Despite its reputation, it was only once successfully attacked, by the Scottish King James IV on his way to the fateful battle of Flodden.

Berwick Barracks Museums Complex.
The Parade, Berwick-upon-Tweed. One of the earliest purpose-built Barracks in the country now houses 3 museums including "By Beat of Drum" chronicling the history of the British Army, the Berwick Borough Art Gallery and the Kings Own Scottish Borderers Regimental Museum. The Art Gallery includes a significant portion of the Burrell collection, more usually associated with Glasgow. At Berwick, the collection includes paintings by Degas and Japanese "Arita" pottery. "A Window on Berwick" is a dramatic reconstruction of life in bygone Berwick.

Belsay Hall Belsay.
A remarkable estate with 14th century castle, 17th century manor house and 19th century neoclassical hall set in 30 acres of landscaped gardens and grounds. The hall was designed by the legendary Newcastle architect, John Dobson, and is exactly 100ft (30m) square. Its present unfurnished state gives even greater emphasis to its bold architecture which is now regularly used to display contemporary artworks.

MINE OF EXCITEMENT
KILLHOPE - A GREAT DAY OUT IN THE COUNTRY

It's dark. It's wet.

It's an adventure!

A trip down Park Level Mine at Killhope, the North of England Lead Mining Museum, is only one exciting part of this exceptional family day out, to the award-winning museum set in the heart of the beautiful North Pennines countryside.

The newly extended mine trip takes you down the original tunnel to discover the working conditions of Victorian miners, your guided tour takes about an hour. We kit you out in hard hat, miner's lamp and wellingtons. On the surface you can see the giant working Armstrong water wheel, which was fully restored in 1991.

Discover how miners lived and worked in the mineshop, dress up in Victorian clothes, don't forget to bring your camera. Discover how boys as young as 10 years old were employed to work as washer boys separating lead ore from other unwanted minerals using hand tools on the washing floor, any 'finds' here are yours to keep.

A new woodland path, suitable for pushchairs takes you around the reservoirs, which are teeming with wildlife. A new hide in the woodland gives you a unique opportunity to observe the endangered red squirrel colony at close quarters. The longer woodland walk incorporates more of the history of lead mining including hushes and shafts. There's also a playground for younger children, especially the under 4's who are too young to go down the mine, but they can get on site free.

Killhope now has a permanent exhibition of breathtaking 'Pennine jewels' - amazing coloured crystals of Flourspar and other minerals unearthed by miners digging for lead ore. Alongside the 'jewels' is a spectacular display of spar boxes made by miners on long, dark nights to show off the dazzling crystals. One spar box is over 7ft tall!

Killhope is an English Tourism Council Quality Assured Visitor Attraction set 1,500ft above sea-level - England's highest tourist attraction. The mine is cool even in the height of summer so wear something warm.

For further information contact Killhope on 01388 537505, visit the website at www.durham.gov.uk/killhope or email Killhope@durham.gov.uk

Opening Times: Killhope is open daily between April 1st to September 30th from 10.30am to 5.00pm, and between July 21st and September 1st until 5.30pm

Week-ends & Half-term in October, and Santa visits us in his grotto (down the mine) on 6th, 7th, 13th & 14th December.

DLI MUSEUM &
DURHAM ART GALLERY

Open every day [bar Christmas Day] April - October 10am - 5pm, November to March 10am - 4pm.

Admission charges.

Location: Durham City, 1/2 mile (800 metres) north-west of city centre off A691, near railway station.

Parking: Large free car park 80 yards (73 metres) from entrance.

Catering: Licensed café with fine views over parkland, serving hot and cold light meals and refreshments.

Set in superb landscaped grounds, the DLI Museum & Durham Art Gallery is one of Durham's most rewarding visitor attractions. Downstairs, the Museum tells the proud story of County Durham's own Regiment The Durham Light Infantry from 1758 to 1968, with particular emphasis on WW1 & WW2. Here the displays focus on the experience of war, using letter & diary extracts, plus the actual voices of DLI WW2 soldiers. Family friendly, there are red tunics for younger children to wear and two Horrible History areas for older children.

Meanwhile upstairs, the Art Gallery presents throughout the year an exciting exhibition programme and regular events from concerts to practical workshops. Past exhibitions have ranged from Beryl Cook's paintings, to Henry Moore's sculptures, to Judge Dredd cartoons. Contact us for a free Events brochure to find out what is on show now.

Outside in the landscaped grounds - ideal for picnics - there is a Guided Walk that rewards visitors with a stunning view of Durham Cathedral..

Disabled Visitors: The DLI is fully accessible to all visitors, including those in wheelchairs. Dedicated parking near entrance.

For further information: Tel: 0191 3842214, Fax: 0191 3861770

Website: www.durham.gov.uk/dli

E-mail: dli@durham.gov.uk

DURHAM COUNTY COUNCIL

BINCHESTER
ROMAN FORT

Binchester Roman Fort

1$^{1}/_{2}$ miles north of Bishop Auckland, County Durham. Binchester, once the largest Roman fort in County Durham, has a rich and exciting history, of Roman legions, pagan Saxons and early Christian faith. In a defensive position overlooking a loop of the River Wear and dating from around AD79, it was one of a network of forts along Dere Street, the main Roman supply route between York and the Firth of Forth.

The fort has survived well, despite stone having been raided for nearby buildings - the Saxon church at nearby Escomb has carved Roman stonework built into its walls! Defensive ramparts are still visible, and green fields hide the buried civilian settlement or vicus. The excavated remains that you can visit lie in the centre of the fort, and include a stretch of Dere Street itself, part of the commanding officer's house, and his private bath-house (the finest example of a military bath-suite in Britain).

There is always plenty to see and do at Binchester - follow in the steps of Roman legionaries, find the carved beast of Binchester, and look for the footprints of a Roman child. Use our trails around the site and find out about Roman life from indoor displays. A

programme of special events throughout the summer recreates history on the site. Binchester has good access and facilities for visitors in wheelchairs, and free parking. The site is open seasonally (please enquire for details). Admission charges apply.
For further information please contact Binchester Roman Fort 01388 663089
Outside opening hours telephone 0191 383 4212
Website: www.durham.gov.uk/binchester

SEATON DELAVAL HALL

SEATON SLUICE, WHITLEY BAY, NORTHUMBERLAND NE26 4QR

The home of Lord and Lady Hastings, half a mile from Seaton Sluice, is the last and most sensational mansion designed by Sir John Vanbrugh, builder of Blenheim Palace and Castle Howard. It was erected 1718-1728 and comprises a high turreted block flanked by arcaded wings which form a vast forecourt.

The centre block was gutted by fire in 1822, but was partially restored in 1862 and again in 1959-1962 and 1999-2000. The remarkable staircases are a visual delight, and the two surviving rooms are filled with family pictures and photographs and royal seals spanning three centuries as well as various archives. This building is used frequently for concerts and charitable functions. The East Wing contains immense stables in ashlar stone of breathtaking proportions. Nearby the coach house with farm and passenger vehicles, fully documented, and the restored ice house with explanatory sketch and description. There are beautiful gardens with herbaceous borders, rose garden, rhododendrons, azaleas, laburnum walk, statues, and a spectacular parterre by internationally famous Jim Russell, also a unique Norman Church.

Open: May & August Bank Holiday Monday: June - 30 Sept: Wed & Sun 2-6pm. Admission: Adult £3, Child £1, Conc. £2.50. Groups (20+): Adult £2.50, Child £1.00, Student £1.00
Tel: 0191 2371493/0191 2370786

RABY CASTLE

STAINDROP - DARLINGTON - CO. DURHAM - DL2 3AH - TEL: 01833 660202
www.rabycastle.com

The magnificent Raby Castle, set in the beautiful North Pennines has been home to Lord Barnard's family since 1626, when it was purchased by his ancestor, Sir Henry Vane, the eminent Statesman and Politician. The Castle was built mainly in the 14th century by the Nevill family on a site of an earlier Manor House.

They continued to live at Raby until 1569 when, after the failure of the Rising of the North, the Castle and its land were forfeited to the Crown.

Today, Raby Castle, home of the 11th Lord Barnard welcomes thousands of visitors each year. Although Raby Castle is a defended home rather than a fortress, it has seen action in battle, notably during the Civil War when as a Parliamentary Stronghold, it was besieged 5 times by Royalists. Fortunately the Castle suffered little damage and it was not until the 18th century that the first major alterations were made to the mediaeval structure.

Every room in Raby Castle, from the magnificent Barons' Hall, where 700 knights gathered to plot the Rising of the North, to the Mediaeval Kitchen which was used until 1954 gives an insight to life throughout the ages.

There is an outstanding collection of arms, armour and sporting trophies, which reflect the interests of many of the current Lord Barnard's ancestors.

The Octagon Drawing Room is by far the most luxurious room at Raby, being designed solely to impress. The Castle is situated amidst a 200 acre Deer-Park where Red and Fallow Deer graze. The beautiful walled gardens with formal lawns, ornamental pond and rose gardens are bound by the ancient yew hedges, an original feature dating back to the 18th century.

The 18th century stable block contains a horse-drawn carriage collection. The State Coach, used at the Coronation of King Edward VII in 1902 is one of these fine examples of travel in bygone days. In the Tack Room are displays of pristine harnesses and trappings.

The 18th century stables have been converted into tearooms, where the old stalls have been incorporated to create an atmospheric setting.

The Woodland Adventure Playground is close to the picnic area.

HIGH FORCE

RABY ESTATES OFFICE - MIDDLETON-IN-TEESDALE - COUNTY DURHAM
TEL: 01833 640209 - FAX: 01833 640963
E-MAIL: teesdaleestate@rabycastle.com www.rabycastle.com

From its rise as a trickle, high on the heather covered fells at the top of the North Pennines, to the top of the whin sill rock at Forest-in-Teesdale, the River Tees steadily grows and gathers pace. High Force, England's largest waterfall then suddenly and spectacularly drops 70 feet into the plunge pool below. The woodland walk leads you to this spectacular sight. As you begin the descent down the gentle slope the well-maintained path twists and turns giving a different view every few yards. The muffled rumble suddenly turns to a roar and the sight astounds you. High Force commands your respect, its power is its beauty, but it must be treated with care and children should be supervised at all times.

High Force, set in England's unspolit North Pennines, with its breathtaking scenery is open for your enjoyment all year. The waterfall is spectacular throughout the seasons. Enjoy a relaxing lunch in the picnic area and visit the gift shop with lovely mementos that will remind you of your visit to High Force.

Situated alongside the B6277, 4.5 miles from Middleton-in-Teesdale approximately 35 minutes away from Raby Castle.

Openings: High Force is open to the public all year round. The falls are manned from Easter to the end of October, weekends and school holidays 10am to 5pm (Due to adverse weather conditions, part of the falls may not always be accessible)

Admissions: Charges apply for the parking of vehicles and admission to the waterfall.

Group Travel: Discount given upon pre-paid bookings. Coaches park free unless parties not visiting the falls but using the facilities. In that case there is a £20 charge.

Facilities: Toilets and picnic areas.

BAMBURGH
CASTLE

The finest castle in England

Standing on a rocky outcrop overlooking miles of beautiful sandy beach, Bamburgh Castle dominates the Northumbrian landscape. The castle became the passion of the 1st Baron Armstrong, engineer and industrialist, who, in the 1890s, began its renovation and refurbishment.

This love of Bamburgh was passed down through the family to the late Lord Armstrong, who personally oversaw the completion of his ancestor's dream. Today, Bamburgh Castle is still the home of the Armstrong family, and visitors are able

work of the 1st Lord Armstrong. An inventive engineer, shipbuilder and industrialist, Armstrong left a great legacy to the modern age and Tyneside in particular. It also includes a fascinating collection of World War II aviation artifacts.

Open
16 March to 31 October inclusive.
Daily from 11am to 5pm (last entry 4.30pm)
Car park opens at 10.30am
Guiding services by arrangement for parties (minimum of twelve visitors)

to enjoy what has been described as the finest castle in all England.

The public tour includes the magnificent King's Hall, the Cross Hall, reception rooms, the Bakehouse and Victorian Scullery, as well as the Armoury and Dungeon. Throughout, these rooms contain a wide range of fine china, porcelain and glassware, together with paintings, furniture, tapestries, arms and armour.

The Armstrong Museum
Occupying the former Laundry Building, the museum is dedicated to the life and

ALNWICK CASTLE

Home of the Duke of Northumberland

Alnwick, Northumberland NE66 1NQ Tel. 01665 510 777 Fax. 01665 510 876
www.alnwickcastle.com

Alnwick Castle has been owned by the Percy family since 1309. Built originally as a defence during the medieval wars with Scotland, it has also been used as a prison in the English Civil war, and from 1949-77 it accommodated a teacher training college.

After its use as a fortress declined, the Castle fell into a state of disrepair until the 1750s, when the Earl of Northumberland (the title of Duke was given in 1766) and his wife decided to turn the castle into a family home. Architects who worked on the restoration included James Paine and Robert Adam.

Thomas Call and 'Capability' Brown landscaped the surrounding countryside and created the stunning views you see today from the Castle windows and gun terrace.

Algernon, the Fourth Duke, who was interested in arts, science, music and the Italian crafts, replaced the gothic style of this 18th century work with more serious architecture. Between 1854 and 1867, under the supervision of Anthony Salvin, hundreds of workers changed the Castle's exterior and interior, with craftsmen brought from Italy to train the local people in the skills of intricate wood carving. These are revealed in the beautiful State Rooms you will visit.

More recently the Castle has been the location for major films, including Harry Potter, Elizabeth and Robin Hood, Prince of Thieves.

Wander around the grounds and marvel at the magnificence of this famous Castle, set in the splendid Northumberland countryside, near the historic market town of Alnwick.

For further details please telephone 01665 510 777 or visit our website.

CHILLINGHAM CASTLE

Northumberland NE66 5NF Tel: 01668 215359

Chillingham Castle is one of the most important pieces of fortified domestic architecture in the county. In the 13th century a mansion tower existed, and here Henry III stayed in 1255 on his return from the borders, while Edward I was here in 1298 on his way to Scotland. In 1344 Sir Thomas de Heton obtained a license to fortify his Mansion of Chillingham with a wall of stone and lime and to convert it into a castle or fortress. The work was completed in 1348.

In the reign of Elizabeth I an extensive reconstruction was planned and the main entrance was moved to its present position in the centre of the no longer hostile north front.

Since Elizabethan times alteration has tended to take the form of adapting rather than rebuilding. The result is that the old remains behind the new and Chillingham is a house of secrets, some of which are only now yielding to patient research. Old stairways have been found mounting the deep wells of the southern towers. The original floor of the solar has been traced behind the old hall on the east. Windows and fireplaces, long obscured behind plaster, have been retrieved, and one walled-in Tudor fireplace has been found to contain over 100 documents, letters and the oldest writ in Northumberland, dating from 1540. Some of these documents can be seen in the museum,.

Since 1933 the Castle has been uninhabited and totally neglected. However the property now belongs to Sir Humphry Wakefield Bart who is presently restoring the castle to its former glory. This massive undertaking has been going on for several years and it is hoped the project will be completed in the next few years. Enjoy your visit and return to see our continuing progress.

Ford & Etal

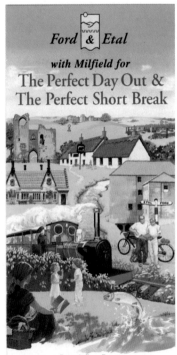

Ford & Etal

with Milfield for

The Perfect Day Out & The Perfect Short Break

FORD & ETAL ESTATES
NEAR BERWICK-UPON-TWEED · NORTHUMBERLAND

A 6,000 hectare rural estate set between the Cheviot Hills and the coast whose attractions include:

- A fully restored water driven corn mill with Granary Cafe and craft shop.

- A narrow gauge steam railway with gift shop and Railway Society room.

- Victorian murals and watercolours by Louisa Anne, Marchioness of Waterford at the Lady Waterford Hall, Ford Village.

- Historic Ford castle with self catering facilities. Guided tours and afternoon teas for groups - by appointment.

- The 14th century Etal Castle with award winning Border warfare and battle of Flodden exhibition (English Heritage). The Black Bull, Etal - Northumberland's only thatched roofed pub.

- B&B and self catering outlets, tea rooms at Ford and Etal, plant nurseries, craft workshops, walking, cycling (cycle hire available) and fishing.

The Perfect Day Out

Battles in Northumbria

The 'border' region between England and Scotland is probably the most fought over land in the United Kingdom. For centuries the area from the River Tweed to the River Tyne has been hotly contested by many different races. Scots, Romans, Danes, Anglo-Saxons and the English have all laid claim to this land, and all prepared to spill blood to keep it. Even the legacy of the Romans in the form of Hadrian's Wall was not enough to prevent centuries of border skirmishes and full scale battles.

To stand in the midst of the Northumberland moorland, with its stark, majestic beauty it is hard to imagine the sounds of steel against steel, axes beating on wooden shields, or the thunder of hooves as mounted warriors rode into the malestrom of a heated battle. It is even harder to imagine the sounds of the wounded and the dying as the battle subsided.

The Anglo Saxon occupiers of England had completed the process of driving the Celts out of the country by 613. Thereafter the country was divided into seven kingdoms: Kent, Sussex, Essex, East Anglia, Wessex, Mercia and Northumbria, collectively known as The Heptarchy. All were intent on establishing supremacy over, or defending themselves from their neighbours. Some, notably Kent and Northumbria fought to convert their neighbours to Christianity. Northumbria was first to gain the upper hand and then Mercia. However, with all the in-fighting going on it left the door wide open for the Vikings to walk in and loot the countryside at their leisure.

Around the year 870 these northern invaders ruled Northumbria from their capital inside the old Roman walls of York. The last Viking King of Northumbria was Eric Bloodaxe, who had been King of Norway in the 930s but was expelled for his extreme cruelty: he is said to have murdered his seven brothers. Eric sailed to England and found the area of Northumberland so much to his liking that he established himself as King of Northumbria. His cruelty continued and he was deposed by Eadred, King of England, he was finally killed during a fierce battle when he challenged Eadred and his army in an effort to regain his lost kingdom. This was the battle that ended Northumbrian independence.

The Battle of Carnham - 1018

This was a highly significant battle where the Scots, lead by Malcolm II King of Alba in alliance with Owen King of Strathclyde regained the Lothian region which had been lost in 1006. Owen had died either at this battle or shortly afterwards and Malcolm made his grandson Duncan King of Strathclyde. Not happy with the news of the defeat at Carnham King Canute summoned the Earl of Northumbria to his court and had him assassinated. The Earl

The silver penny of Eric Bloodaxe bears an unsheathed sword, a symbol of his heathen rule in Northumbria.

was succeeded by his brother Eadulf and in return for peace he ceded Lothian to Malcolm.

The Battle of Alnwick - 1093

This battle was as a result of a raid on England by Malcolm III King of Scots after William II of England had extended control over Cumbria and fortified Carlisle. Nicknamed 'Rufus' because of his fiery red hair, William was reputedly intelligent, witty and generous to his soldiers. He was also cynical and ruthless, not a man to be crossed. He was not prepared to back down in the face of threats from Malcolm. A battle was inevitable. The Scots were soundly beaten and Malcolm died in the fighting. In 1774 Malcolm's cross was erected to mark the spot where he fell at the hands of Robert de Mowbray, Earl of Northumberland.

Halidon Hill - 1333

In the early part of the 14th century, Berwick was a thriving trading town, named at one time as the Alexandria of the north. It's prime location being by the sea and on the Scottish, English, border meant it was a desirable place to have as part of your country.

By 1333, Berwick belonged yet again to the Scots and the English decided it was about time they got it back. Edward III, who was aged 20 or 21 at the time moved north with his army to reclaim Berwick. Douglas, a Scot, decided to repeat an earlier manoeuvre to distract the English by sending troops to Bamburgh where he had heard the English Queen was being accommodated. Alas, Edward III was made of stronger stuff and held his position, not returning to Bamburgh to protect his wife.

A treaty of surrender was signed and Berwick was to be handed over to the English by July 21st. In the mean time there was to be a truce and hostages were exchanged. On July 19th Berwick was put up as a winning prize after a battle. The English, wisely, positioned themselves on Halidon Hill which left the Scots with the undesirable position of having to fight uphill. It was a disaster for the Scots who suffered many casualties. Douglas, the leader of the Scots, died with his men along with many other nobles. The English had very few casualties due to their prime position and well equipped army. Berwick fell once again into the hands of the English - but not for the last time!

Otterburn - 1388

Here a moonlight battle was fought between "Harry Hotspur", Sir Henry Percy and his long-standing enemy James, Earl of Douglas. The battle went long into the night and 'Hotspur' was captured. The Scots won this battle but their leader was killed during the fighting.

Humbleton Hill - 1402

After his release Sir Henry sought revenge and defeated the new Earl Douglas at this battle on a rounded hill near Wooler. Sir Henry later died at the battle of Shrewsbury and a cross was erected near Otterburn as a memorial.

Battle of Hedgeley Moor - 1464

This battle was fought during the English civil war and involved the House of Lancaster and the House of York. Sir Ralph Percy, was defeated by the House of York in this war of the roses.

Hexham - 1464

The battle of Hexham on 15th May 1464 brought to a close the phase of the Wars of the Roses that had begun in 1459 at Ludford Bridge. Its result saw the death, imprisonment, exile or reconciliation of all the leading Lancastrians and left Edward IV in undisputed control of the kingdom, with the aid of the Earl of Warwick amongst others. This happy state of affairs was to continue until the falling out between

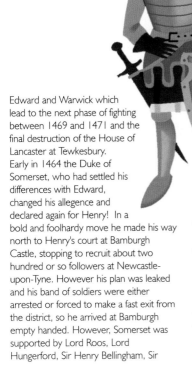

Edward and Warwick which lead to the next phase of fighting between 1469 and 1471 and the final destruction of the House of Lancaster at Tewkesbury. Early in 1464 the Duke of Somerset, who had settled his differences with Edward, changed his allegence and declared again for Henry! In a bold and foolhardy move he made his way north to Henry's court at Bamburgh Castle, stopping to recruit about two hundred or so followers at Newcastle-upon-Tyne. However his plan was leaked and his band of soldiers were either arrested or forced to make a fast exit from the district, so he arrived at Bamburgh empty handed. However, Somerset was supported by Lord Roos, Lord Hungerford, Sir Henry Bellingham, Sir

Ralph Percy and the former Yorkist Sir Ralph Grey, together they built a reasonable force that was to fight one of the most decisive battles in the history of England at which the House of Lancaster was victorious.

The Battle of Flodden - 1513

King James IV of Scotland was defeated with the majority of noblemen this resulted in the end of Scotland as a separate country. King James had agreed to attack England as an attempt to divert troops away from their French campaign. This was the last and most bloodiest battle fought in Northumberland. The Scots King and most of his noblemen were killed during this battle.

Welcome to Northumberland

Northumberland, England's border Country stretches from Seaton Delaval in the South to the borders of Scotland, Berwick-upon-Tweed, in the North - 60 miles in total. It has much to offer visitors from its beautiful countryside, with the rolling Cheviot Hills and the Northumberland National Park, to its stunning coastline, most of which has been designated as an Area of Outstanding Natural Beauty. Northumberland has one of the cleanest coastlines in the country. In 2001, its 12 main beaches all passed the Environment Agency's basic standard for Coastal Bathing Water and six beaches received the Rural Seaside Award for their cleanliness and safety.

The County boasts of its golden sandy beaches, open countryside and the leafy forest that surrounds the largest man-made Lake in Europe, Kielder-Water reservoir. Northumberland has a greater concentration of castles open to the public than anywhere in England, some of the most dramatic are Bamburgh, Dunstanburgh and Warkworth. Bamburgh rises magnificently from the smooth golden sands and has provided the backdrop to countless films. Further inland are some of the finest castles and stately homes, often used in major television and film productions. Alnwick Castle, home to the Duke and Duchess of Northumberland, was magically transformed into Hogwart's

School of Wizardry in the blockbuster film Harry Potter and the Philosopher's Stone. Picturesquely located to the north of Alnwick town, its austere walls give no hint of the luxurious 19th century interiors within.

No visit or stay in the North of England is complete without a trip to Hadrian's Wall-a World Heritage Site, where a wealth of unique Roman forts and museums almost two thousand years old line the 100km long monument. Some portions of the wall were 15 feet high and punctuated by defensive forts, signal posts and milecastles. If shopping is your interest then there are a number of picturesque market towns such as Rothbury, offering a blend of high street stores as well as a hoard of specialist shops selling Northumberland food such as Northumberland cheese. Antique hunters can obtain a bargain as they browse around the streets and market squares in Corbridge, Hexham, Morpeth, Alnwick and Berwick-upon-Tweed.

Northumberland offers the freshest air you have ever breathed, uncrowded roads and the choice to be as active or laidback as you wish. You may Walk, Ride, Play Golf, Cycle, Fish or simply pamper yourself amidst stunning scenery.

After a long rewarding day it is time to relax over a meal or a drink in one of a number of country pubs, many offering log fires and all offering a warm welcome where no one is a stranger for long. Once visited you'll yearn to return!

Tel: 01670 533924 for a free holiday guide

Eating Out In Northumberland

Northumberland has a host of eating places from fine restaurants, country pubs and cafés to top quality hotels. When eating out, local meat specialities to look out for include tender Cheviot Lamb, which is often served with a redcurrant sauce and chevon (goat meat) served in a goulash style casserole. As there are 5 times as many sheep as people in Northumberland it is not surprising to find this meat on most good hotel and restaurant menus. Venison and other game including grouse is also a local speciality which may or may not be smoked.

Northumberland is also well known as an area for good fishing so, it is not surprising to find fresh salmon and trout on the

menu but there are also many local shellfish delicacies, these include Lindisfarne oysters and langoustines, muscles, crabs and lobsters. Craster is a well-known fishing village and is the home of the world famous oak-smoked Craster Kippers that can be found in starters such as Craster Kipper pate. Some visitors say the traditional English fish and chips and fresh crab sandwiches served in Northumberland are the best they have ever tasted.

Fruit is plentiful in Northumberland especially hedge fruits such as raspberries, blackberries, redcurrants and logan berries which may be served with cream, yoghurt or a contrasting fruit sauce.

Northumberland also grows very high quality grains, especially wheat, barley and oats. Heatherslaw corn mill is a working water mill where visitors can see the production of their own flour and can buy some from the mill shop. The self-raising flour Be-Ro was invented by a local grocer from Longhorsley after he mixed plain flour and baking soda in the correct proportions, it first sold under the name of "Bell's Royal".

There are a wide variety of cheeses made by the Northumberland Cheese Company at Blagdon, near Morpeth, they include Northumberland Cheese itself, Redesdale sheep's cheese, and Elsdon goat cheese. The traditional bread of Northumberland is the stottie - a round, flat, large white bap and this may be accompanied at tea-time by girdle cakes and singing hinnies.

Northumberland also has its own traditional drink - Lindisfare Mead which is made at St Aidan's Winery on Holy Island. It is a sweet, honey based alcoholic drink that was made by the St Aidan Monks. Visitors can sample the Mead at the winery before they buy. There are also three small local breweries whose beers are on sale in local pubs, these are the Border Brewery at Berwick, the Northumberland Brewery at Bedlington and the Hexhamshire Brewery. The famous natural mineral water "Abbey Well" comes from a spring near Morpeth. In addition to serving good local dishes, most restaurants in the County also offer a wide range of international cuisine, including vegetarian options. In all the eating establishments throughout the County you will receive a well-prepared meal served in warm, relaxing surroundings with welcoming hosts.

Gardens In Northumberland

William Turner, Father of English Botany, was educated in the 16th century at Morpeth Grammar School that is now the Chantry Bagpipe Museum. His work is commemorated in the medieval garden within Carlisle Park, Morpeth. William wrote the first Herbal, a collection of plants names in Latin. Medieval instructions can be found in the Park's walled garden. Wallington Hall dates from 1688 and includes a walled garden and woodland walks by ornamental ponds which visitors can enjoy.

Belsay Hall features 30 acres of magnificent gardens, a mix of formal and informal with most created in the 19th century. Here you will find the famous quarry gardens with their distinct micro-climate. Closer to the house are the terraced gardens with deep mixed borders, a magnolia terrace, and rhododendrons may be found in bloom even in the middle of winter. The sunken lawns are used for croquet

Many famous gardeners have left their unique mark upon Northumberland when they have created beautiful landscapes throughout history in the large country houses of the County. Lancelot "Capability" Brown (so named because of his regular advice to clients that their land had "capabilities") was born at Kirkharle near Morpeth in 1716.

Lancelot returned to Northumberland after moving south to find fame and fortune to work for Sir William Blackett of Wallington where he landscaped Rothley Lakes. His work can also be seen from Alnwick Castle in their grounds. Visitors to Northumberland can see an exhibition of Capability Brown's work at Kirkharle Courtyard.

tournaments. The Cragwood Walk is especially popular in February when it has a carpet of snowdrops.

Cragside is famous for their early summer displays of colourful rhododendrons. The estate boasts of 900 acres of hillside, lakes and woodlands with paths and country drives. Cragside House was the first house in the world to be lit by hydroelectricity and "The Power Circuit" is a 1.5 mile circular walk, which includes the restored Pump and Power Houses with their hydraulic and hydro-electric machinery. The remarkable Orchard House, ferneries, rose loggia and Italian gardens are all within walking distance of the house.

The Alnwick Garden is probably the most recent garden to be opened to the public in Northumberland.

The site is magical, the designers world class and the vision was clear - to create one of the world's greatest contemporary gardens here in Northumberland.

It was conceived as a unique and modern garden, designed to be relevant to future generations. The Alnwick Garden is not simply a recreation of the past - it is a fresh approach to design. Use of the latest technology has created a contemporary composition of ever-changing sounds, sights, textures and smells.

It is an architectural garden, designed to look attractive all year round, and is dominated by the movement of water. Cascades, waterfalls, fountains, waterspouts, and pools create ambiences to startle, soothe, delight and amaze visitors.

Tillmouth Park
— COUNTRY HOUSE HOTEL —

Tillmouth Park is named after the river Till & the 15 acres of parkland in which it is set. It is a magnificent 1882 English mansion house on the Scottish Border near to the famous River Tweed. Tillmouth Park holds a 3 Star Silver Quality Award from the English Tourism Council and boasts AA Rosette awarded food. All of the fourteen ensuite bedrooms are individually designed with period and antique furniture and they are fully appointed with tea-making tray, hairdryer, trouser press, bathrobes & toiletries. Three of the bedrooms have four poster beds, whilst all State and Deluxe rooms have large picture windows and superb views to the surrounding countryside. The wood-panelled Library Dining Room offers contemporary British cuisine featuring locally sourced specialities and there is a Private Dining Room featuring a marble fireplace. The Park Bistro offers a more informal menu and is situated next to the well-stocked bar presenting a wide range of malt whiskies. The galleried Main Hall features wood panelling and stained glass doors and the drawing room offers superb elegance, with open log fires throughout the public rooms. Service bells are still available to

Your Personal Country House

If you long for peace and tranquillity away from everything, you can hire Tillmouth Park as your own personal retreat. Take advantage of the magnificent reception rooms and hire the whole hotel for your special occasion. The exclusive use hire covers a twenty four hour period and includes all fourteen bedrooms on a bed and breakfast basis and private use of the rest of the hotel and its grounds. Even before you arrive, all your transfer requirements from Edinburgh or Newcastle airports or from Berwick

summons staff if needed.

Tillmouth Park is your perfect home in the country. It is ideally situated for country pursuits with fishing on the Tweed & Till. Clay shooting is available within the grounds, whilst rough & game shooting is easily arranged locally. The area also has many fine challenging golf courses. Exclusive use functions, conferences and weddings are a speciality at Tillmouth Park.

mainline railway station can be taken care of. If it's a self-drive 4 x 4 or a chauffeur driven limousine, put yourself into the hands of your staff at Tillmouth Park and leave it to them.

Tillmouth Park Country House Hotel,
Cornhill-on-Tweed,
Northumberland.
TD12 4UU
Tel: 01890 882255
Fax: 01890 883175
info@tillmouthpark.f9.co.uk

Hadrian's Wall World Heritage Site

Looking east from Winshields - Graeme Peacock

Hadrian's Wall alone would make the area special, but there is so much more to this wonderful part of Britain. On either side of the famous Wall the countryside has a real frontier feeling - it is not just the northern limit of the Roman Empire, but the scene of ancient battles between the English and the Scots, regular raids by sixteenth century outlaws the Reivers, and gatherings of Jacobite rebels.

Today it is tranquil and friendly. Alongside the rich historical heritage are modern facilities and a wide variety of attractions. The World Heritage Site spans northern England, from the salt marshes of the Solway coast in Cumbria to the sandy beaches east of Newcastle. Scenery ranges from infinite to the intimate; is classic walking country, quiet cycling country and offers superb sightseeing and shopping.

Hadrian's Wall itself can be seen at various locations. Short sections appear beside the pavements in Newcastle and in fields in the Tyne Valley, but the most spectacular stretch is between Birdoswald in Cumbria and Housesteads in Northumberland. Museums and forts along the 73 mile length of the Wall use innovative displays and reconstructions to tell a fascinating story.

Archaeologists are making exciting discoveries all the time and every aspect of Roman life is on view. Patrolling soldiers must have been grateful for the shelter of the milecastles built into Hadrian's Wall, and even more so for the relative luxury of their barracks at the forts. Commanding Officers lived in some style, as can be seen at Arbeia Roman Fort in South Shields, whilst a working Roman bath house can be experienced at Segedunum across the River Tyne in North Shields. You will feel very close to the former occupants of Vindolanda Fort after seeing their personal possessions, excavated in almost perfect condition.

Many of the sites can be reached by public transport, using Tyne Valley Line Trains, the Hadrian's Wall Bus (service AD122), scheduled bus service or the Tyne and Wear Metro.

Reconstructed Roman Temple - Graeme Peacock

Birdoswald Roman Site - Philip Nixon

Sycamore Gap - Philip Nixon

Cawfields Milecastle No. 42 - Graeme Peacock

The countryside in the World Heritage Site is stunning. Walkers can enjoy a network of paths and in summer 2003 the Hadrian's Wall Path National Trail will open along the entire length of the monument. There are already paths alongside the best preserved sections that offer views across Northumberland National Park to the Cheviot Hills in the north and to the Pennines in the south. This is the land of the far horizon.

Cyclists rave about the area as well. Quiet lanes and farm tracks are ideal for a family day out, while rugged off-road routes challenge the more adventurous. The Hadrian's Wall Bus and Tyne Valley Line trains carry bikes. The long-distance Hadrian's Cycleway will be fully opened in summer 2003.

Other attractions include two great cities.

Carlisle and Newcastle, historic houses and gardens, castles and towns and villages with some great shops, local markets, fairs and special events. Their pubs and tea rooms will encourage you to relax and linger as you contemplate the pleasure of following in Hadrian's footsteps.

New for May 2003 will see the opening of the Hadrian's Wall Path National Trail. The 81 mile route roughly follows the line of Hadrian's Wall, stretching across northern England from Bowness-on-Solway on the west coast to Wallsend, near Newcastle upon Tyne on the east coast. The Path, which will be the only National Trail to lie within a World Heritage Site, will take walkers through stunning scenery and historic landscape.

See archaeology in action in Vindolanda Fort and museum. From April to the end

Vindolanda Roman Temple - Philip Nixon

Birdoswald - Philip Nixon

Cuddy's Crag - Keith Paisley/National Trust

of August, Sunday to Friday there are excavations in progress. Watch finds emerge and don't miss the chance to chat to archaeologists.

Every footstep counts, taking care of the Wall.

During the winter months the World Heritage Site is an especially fragile environment.

You can help protect one of the great wonders of the world by following the advice below.

* Always keep to signed paths.
* Visit the organised paying sites, which are more robust and can accommodate visitors, but please avoid walking alongside the Wall when the ground is very wet.
* Please walk beside the wall and not on it.
* Respect livestock and land.
* Keep dogs on a lead.
* Use public transport whenever you can.

Walltown Quarry - Tony Hopkins

Vindolanda - Chesterholm Museum - Philip Nixon

For further information telephone the
Hadrian's Wall Information line on
01434 322002
or email **info@hadrians-wall.org**
Postal enquiries to
Hadrian's Wall Information,
Station Road,
Haltwhistle,
Northumberland,
NE49 9HN.

East from Walltown - Graeme Peacock

The Museum of Antiquities

University of Newcastle-upon-Tyne - Telephone: 0191 2227849

As the main museum for the World Heritage Site of Hadrian's Wall, Newcastle's Museum of Antiquities' spectacular Roman collection includes material drawn from the entire Military Zone, which was the Roman Empire's most north-westerly frontier for nearly 400 years. The collection includes artifacts from almost every province of the Roman Empire and the military units and civilian populations of these provinces. A large collection of Roman

inscriptions and sculptural stones includes many important pieces such as the Birth of Mithras stone from Housesteads, which has the earliest depiction of the signs of the Zodiac anywhere in Britain. The Roman jewellery assemblage is one of the most comprehensive in Britain and boasts internationally acknowledged masterpieces, such as the Aesica Brooch and Aemilia Finger Ring, thought to be the earliest Christian artefact yet found in Britain. Other material includes many unique examples of the full range of Roman domestic artifacts known in the north-west provinces.

Although best known for its Roman collection, the Museum's Prehistoric and Anglo-Saxon collections are also internationally renowned. The former includes the most significant collection of prehistoric rock art in Britain, notable groups of Neolithic stone axes, Bronze Age pottery and metalwork. The Anglo-Saxon exhibit features examples of almost all the monuments known from the period. Significant pieces include the Rothbury Cross, which is the earliest Christian rood surviving in the country and the Alnmouth Cross, which bears the only attested Irish name on an Anglo-Saxon artefact.

The Museum also has a high level of achievement in research, publication and conservation and these, together with the quality of its collections, make it one of the finest in the world.

Opening Times:
Mon - Sat 10.00am - 5.00pm
Sunday - Closed
24 - 26 Dec. - Closed
Jan 1st & Good Friday - Closed

THE COUNTY HOTEL

Priestpopple - Hexham - Norhtumberland - Tel: 01434 603601

Situated in the centre of Hexham, an Abbey Town in the heart of Hadrian's Wall Country, The County Hotel offers a restful corner behind its old fashioned revolving door, to ease tired feet, and rest a tired body.

The small, family run hotel is proud of the variety of facilities it has on offer, all of which reflect the meaning of hospitality.

Recently refurbished, "The County" offers a variety of bedrooms, including four poster beds, king sized beds, family rooms and room dedicated to the single traveller. All have private facilities, refreshment tray, television and telephone.

From the friendly greeting to the cheerful goodbye, the Harding family wish to make your stay as comfortable and enjoyable as possible.

"The County" is also taking great pride on

the reputation it is gaining for wholesome, freshly prepared food offered in the Restaurant and other areas, prepared from locally produced and sourced ingredients. The menu offers traditional dishes, produced with care and attention.

The bars at "The County" serve a variety of Real Ales and Wines from around the world, in areas which continue the overall theme of relaxed and quiet enjoyment, providing the ideal venue to meet friends and relax before dining, or exploring the area.

Whilst not having a dedicated car park of its own, there are several within easy reach, and "on street" parking is allowed in Priestpopple.

The Harding family will gladly assist in providing details of local knowledge in relation to walks, attractions and events of interest to visitors and tourists, and enjoy hearing of their guests experience of Northumberland.

The hotel is housed in a listed building of some age, the layout of which does not lend itself easily to the installation of a lift. Nevertheless, the stairs are shallow, and the Hardings will do their best to accommodate individual requests for easier access, made in advance.

Rye Hill Farm

Slaley, Nr. Hexham, Northumberland NE47 0AH
Tel/Fax: 01434 673259
e-mail: info@ryehillfarm.co.uk Web: www.ryehillfarm.co.uk

Rye Hill Farm offers you the freedom to enjoy the pleasures of Northumberland throughout the year while living comfortably in the pleasant family atmosphere of a cosy farmhouse adapted especially to receive holidaymakers, both bed and breakfast and self-catering. Well-mannered children and pets are more than welcome.

This is a 300 year old, Northumbrian, stone farmhouse - set in its own 30 acres of rural Tynedale. The self-contained accommodation in the converted farm buildings offer all the comforts of good farm hospitality. A network of footpaths provide ready access to the countryside. There are interesting walks along riversides, through the nearby forest or up onto the moors. The famous, unspoilt castles, coastline, moors and quiet country lanes of Northumberland, together with Kielder Forest, Kielder Water and the historic City of Durham are all less than an hour's drive

away from Rye Hill Farm - thus making it an ideal base for touring.

Bed & Breakfast - We provide a full English Breakfast and an optional home-made three course evening meal in the dining room, which has an open log fire and a table licence. Our bedrooms are equipped with television and tea & coffee making facilities, and all are en-suite and centrally heated.

Self-Catering - The Old Byre now accommodates up to 9 people self-catering. Ground Floor - suitable for wheelchair users, the living accommodation is open plan with stable-stall kitchen fully equipped with dishwasher, microwave, gas stove and fridge freezer. The lounge-dining area is spacious with a log burning stove, tv and comfy sofa.

Home made evening meals can be ordered in advance from the farmhouse, and all linen and central heating is included in the price.

Log on for the latest up-dates & special offers

Abbey Bistro

45 Hall Stile Bank · Hexham · Northumberland · NE46 3PQ · Tel: 01434 607616

Located on the road into Hexham, opposite Tynedale Retail Park next to O'Neill's Antique Warehouse, The Abbey Bistro has a prime position with ample car parking nearby.

We are a small, intimate but friendly, licensed café serving an extensive menu of hot and cold meals to eat in or take away. Open 7 days a week until around 5.00p.m. we serve all day breakfasts at unbeatable value. We are famous for our wide range of home made soups and quiches and our traditional Sunday lunch is guaranteed to bring you back for more. Our chilled section includes a variety of delicious cakes, tray bakes and scones. We use only the finest teas, including herbal and fruit teas. The best fresh, filter coffee, or, if you prefer a continental taste, a cappuccino, latte or an espresso are available.

Lunch is served all day and we have a daily specials board as well as our regular menu. Our desserts are wickedly tantalising to the taste buds and our portions are not for the faint hearted.

We serve a selection of cold drinks from our chilled cabinet as well as canned beers and lagers. For the more discerning palate, we have a selection of fine red and white wines, both in small and large bottles.

Whatever refreshment you need, visit The Abbey Bistro for friendly service, good food and unbeatable value we can promise you that!

𝔇𝔶𝔳𝔢𝔩𝔰 ℌ𝔬𝔱𝔢𝔩

Station Road, Corbridge, Northumberland NE45 5AY Tel 01434 633633

Corbridge is a pretty residential village, once a market town, four miles east of Hexham on all major routes to Newcastle. A Tyne crossing-place since Roman days, the present bridge has stood since 1674, and was the only one on the river to survive the great flood of 1771.

In Roman times, Corbridge was the main town in the area, it acted as a garrison town for the central sector of Hadrian's Wall, and extensive remains can still be seen at Corbridge Roman Site.

Corbridge, which in Roman times, was the site of Corstopitum, a fort built around A.D. 80, by the Roman Governor of Britain, Julius Agricola. Corstopitum guarded an important crossing of the River Tyne, located at the junction of the two important Roman roads known as Dere Street and the Stanegate.

To the north of Corbridge at a place called Port Gate, the Roman Dere Street crossed Hadrian's Wall, as it continued north into Redesdale on its way to Caledonia.

The intriguingly named 'Dyvels Hotel'! could have taken its name from another Roman road close by known as the `Devil's Causeway'.

This cosy, stone building welcomes visitors with a great atmosphere, personal service and an excellent menu of home-cooked food. The separate restaurant serves a full English breakfast and a wide choice of home-cooked dishes for lunch and dinner. Bar meals are also available to accompany an excellent choice of cask ales or fine wines. With single, double, twin and family rooms available; all tastefully decorated and furnished and equipped with TV and beverage tray, the hotel is an excellent base for touring the district. Hadrian's Wall is just a short drive away, and the surrounding countryside is bursting with first class golf courses such as Slaley Hall, Hexham and Matlan. For those who enjoy a good day out at the races, then Hexham and Newcastle race courses will fit the bill, both are within easy reach of the Hotel. Telephone 01434 633633 for more information.

John Martin Street
Haydon Bridge
Northumberland
NE47 6AB
Tel: 01434 684227
Fax: 01434 688413

The Anchor Hotel stands proud on the south side of the Old Bridge at the entrance to Haydon Bridge and is midway between Newcastle and Carlisle.

First reference to The Anchor can be found in 1422 when its' history began with the hanging of John Badger for felony. Later the hotel was confiscated by the Admiralty and so named The Anchor.

The Haydon Bridge hunt was formed here in 1812 and is understood to be the oldest hunt in Northumberland.

Nowadays, The Anchor enjoys much quieter and relaxed times. The hotel is graded three star by the English Tourist Council and is fully licensed. It is also in the centre of Roman Wall country.

The Anchor is the perfect base for exploring all of the splendid and fascinating attractions that this beautiful part of Northumberland has to offer. Whether it's discovering the Roman Wall, enjoying the peace and tranquillity of Kielder Forest or a day ambling through the many interesting shops in Hexham you are assured to find plenty to do and see.

For your comfort, all of our 10 comfortable ensuite bedrooms have colour television, radio, direct dial telephones and tea/coffee making facilities. There is also a quiet residents lounge for guests which compliments our Riverside dining room.

The Admirals Restaurant enjoys views over the South Tyne River to the centre of Haydon Bridge. The restaurant is fully licensed. After your meal, why not relax by the log fire in the Riverside Lounge. Here you will find a vast selection of ales, spirits and liquors amid local gossip, plus the chance to enjoy one of the many live music nights we hold.

Battlesteads Hotel

Wark - Hexham - Northumberland - NE48 3LS - Tel: 01434 230209 - Fax: 01434 230730
email: info@Battlesteads-Hotel.co.uk - www.Battlesteads-Hotel.co.uk

Situated in the heart of the Northumberland National Park, the Battlesteads Hotel & Restaurant is an ideal location for those wishing to explore the charms of the Northumbrian countryside and its history. Northumbria, England's most northerly region, is unique. Areas of outstanding natural beauty, national parks and spectacular coastlines combine to form Britain's most unspoiled countryside. At the Battlesteads Hotel you'll be staying at the very heart of what is thought of as England's BEST kept secret. With fine food, comfortable surroundings and local activities such as the regular Folk music nights, we're sure you will have a magical and rewarding stay.

The Battlesteads Hotel is a converted 17th century farmhouse, which has been carefully modernised to provide you with comfortable and friendly surroundings while retaining its rural charm and character.

What separates The Battlesteads from the rest of the Hotels in the area is the quality of the food and the dining experience which it has to offer in its Lindisfarne restaurant. The Hotel is justly proud of this wood-panelled 70 seated restaurant. Its plush upholstery and real coal fire provide the perfect setting for business, family or special occasions.

George, the head chef, is renowned for his speciality and al la carte menu which both guests and visitors agree are second to none. For those wishing to experience the traditional Northumbrian Dinner we recommend you visit the restaurant on Sunday afternoon as the Battlesteads Hotel hosts a traditional English carvery. The

carvery is especially popular with the locals and guests alike offering excellent local Cheviot Lamb, Northumberland Beef and local Roast Pork.

Great food needs great wine and as you would expect from a quality restaurant such as ours we have an extensive wine list - but for those wishing to try the local real ale, we at the Battlesteads have our very own Battlesteads Bitter which we can assure you is most enjoyable.

The Battlesteads caters for all occasions both large and small. We pride ourselves on being able to host weddings, christenings, birthdays etc.. with a capacity for 5 to 200 guests. Through the use of an outside marquee we have successfully hosted a number of large weddings. The Battlesteads is also renowned for its regular Sunday music evenings, which attracts people from throughout the area and in

which they participate in an evening of open music where anyone who wishes to play an instrument or sing can do so uninterrupted.

The Battlesteads Double en-suite guest rooms are well-appointed and maintained and allow you to stop worrying about nights of lost sleep and concentrate on enjoying your stay in our hotel.

We have 14 bedrooms, 4 of which are on the ground floor and are fully equipped for disabled people, also one of the rooms has an induction loop for people with hearing difficulties. Disabled parking is next to the main hotel entrance with easy ramp access into the building.

All rooms are ensuite and have colour televisions, ironing boards, trouser-presses and hair dryers.

Each room also comes equipped with tea and coffee making facilities and a choice of continental or Full English Breakfast is served in the morning.

The Battlesteads ensuite family guest rooms encompass quality with economy. If you would like to keep track of the little ones, the family room is an ideal choice as behind a covered wall, a collapsible bunk bed can be lowered allowing two extra occupants to share the room.

Oakey Dene

Allendale, Hexham, Northumberland NE47 9EL
Tel/Fax: 01434 683572

Member of Northumbria Tourist Board
4 Diamond Guest Accommodation Welcome Host Award

Just a short drive from Allendale town is the delightful Oakey Dene Bed and Breakfast. This is the private home of Alan and Carol Davison and they extend a warm welcome to their many guests. Dating back to 1823, this was originally a small cottage before being extended to create this fine, large house. The property has been extensively refurbished, retaining many of the original features, while creating a comfortable home. The rural setting is delightful, and the front of the house is covered with a large creeper and other trailing plants while there is also a colourful garden.

Inside, visitors can make use of a sitting room and dining room. The whole of the interior is kept cosy and warm with open log fires that are lit when the weather turns cold. There are three guest bedrooms in all. The double has an ensuite bathroom, the twin room has a private bathroom and the single room has the use of the bathroom just down the hall.

The delicious home cooked breakfast is served each morning with home made bread and preserves. An evening meal can be provided by prior arrangement. Children are most welcome, maybe they would like to see the prize winning herd of Toggenburg Goats.

Northumberland National Park

©Northumberland National Park Authority. Photographer Simon Fraser

Northumberland National Park is a wonderful place, an awesome landscape dominated by vast skies providing breathtaking views that run on for miles - it is 'the Land of the Far Horizons'. Here you will glimpse wilderness and open landscapes that stir up a deep sense of the past. It is a place where you can find a true sense of tranquillity and escape the stresses and strains of modern life -or you can find adventure! It's up to you. Northumberland National Park is the most northern National Park in England and stretches from the Scottish Border in the north to just south of Hadrian's Wall World Heritage Site, covering an area of more than 1030 sq. kilometres (400 sq. miles). It offers some of the best countryside walking found anywhere in Britain, ranging from short leisurely strolls and picturesque village rambles to more challenging hill and moorland treks. With more than 900 km (600 miles) of way marked footpaths and bridleways to choose from, you will be spoilt for choice!

The upland scenery of the Cheviot massif is second to none with the Cheviot itself rising to 815m (2,674ft). Here and in the Simonside and Harbottle Hills, you can use a range of self-guided leaflets to explore some of the best-preserved prehistoric hillforts in Britain, with place names that evoke a time long gone such as Yeavering Bell, Brough Law and Lordenshaws. The Cheviots also provide some of the best upland stretches of The Pennine Way, Britain's best-known National Trail.

South of the Cheviot Hills, the pastures and river valleys of the Coquet, Rede and North Tyne soften the profile. Hidden amongst this beauty, the remains of medieval castles and bastles (fortified farms) pepper the landscape, conveying a more sinister legacy, that of the Border Reivers. This land in the 15th and 16th centuries was a buffer between warring kingdoms, a wilderness swept by the fear of Reivers. These rustlers and thieves terrorised the Borders, pitting family against family and valley against valley and it marked one of the bloodiest chapters of Northumbrian history. Today, the remains of castles and restored bastles are open to public view. The picturesque village of Rothbury has its own village trail, riverside and local forest walks that satisfy the more casual walker. Many other villages such as Otterburn, Elsdon, Greenhaugh and Falstone are excellent places from which to explore the National Park. Falstone for example lies at the east end of Kielder Water and Forest Park and is conveniently close to all the recreational facilities that the area has to offer (cycling, horse riding, fishing, walking and water-sports).

©Northumberland National Park Authority. Photographer Simon Fraser

At the end of the day, why not drop into a traditional village pub where you may come across live music, a games night or just a quiet evening when you can join in a friendly chat with the locals.

Hadrian's Wall World Heritage Site, Britain's longest monument, stretches for 73 miles across the north of England and the impressive central section lies within the National Park. In May 2003, the new Hadrian's Wall Path National trail will open, enabling walkers to walk from Newcastle-upon-Tyne, on the east coast to Bowness-on-Solway on the west coast, following the line of the wall. The award winning Hadrian's Wall Bus Service runs between Newcastle and Carlisle each summer and stops at most of the Roman attractions along the Wall as well as visiting picturesque market towns and villages along the way. It can carry bikes too!

©Northumberland National Park Authority.
Photographer Simon Fraser

A successful trip starts at home. To get the most from your visit, please make use of our website at www.nnpa.org.uk or the Hadrian's Wall website www.hadrians-wall.org. Numerous guidebooks and literature are available on this treasured landscape such as Walks in Reiver Country containing details of circular walks, maps and route descriptions.

Free leaflets, transport information, location of public car parks along with details on accommodation can be obtained from the National Park Visitor Centres at Rothbury (01669 620887), Ingram (01665 578248) or Once Brewed (01434 344396). Northumberland National Park Authority, Eastburn, Hexham, Northumberland NE46 1BS

©Northumberland National Park Authority.
Photographer Simon Fraser

THE OTTERBURN TOWER

Otterburn - Northumberland - NE19 INS - Tel: 01830 520620

A truly distinctive fortified Northumbrian country house founded by a cousin of William the Conqueror in 1076. It attractively combines almost a thousand

years of history with all the features required by today's discerning guests. The battle of Otterburn took place on the moonlit night of Wednesday 19th August 1388, between the Scots and the English; a truly fierce and bloody conflict with the English suffering heavy casualties. The supposed site where the Battle of Otterburn was fought is just north on the A696 about a mile north west of Otterburn Tower. The position is marked by an old beam from our kitchen and a full account of the battle is given on a notice board at the site.

A former owner and well-known Jacobite sympathiser, Squire 'Mad Jack' Hall, was condemned and executed at Tyburn for treason in 1716. The present owner, Mr. John Goodfellow bought it in August 1998. It was renovated and refurnished during the winter months before it reopened in June 1999. The accommodation is of a standard that compliments the magnificent building. You can rest in the lounge and be warmed by one of the many open log fires, or dine in the wonderful restaurant with the 600 year old oak panelling. The

library has been transformed into a superb bedroom, complete with oak panelling and four-poster bed. You never know, you may experience one of the past guests who is no longer part of our world!! and discover more than you bargained for. The atmosphere truly epitomises all that The Tower has been throughout its coulourful history.

Riverdale Hall Hotel

AA ★★
RAC ★★

Bellingham, near Hexham
Northumberland NE48 2JT
Tel: (01434) 220254
Fax: (01434) 220457
www.riverdalehallhotel.demon.uk

Situated on the outskirts of Bellingham in the Northumberland National Park . The Riverdale Hall Hotel stands in beautiful private grounds on the banks of the North Tyne River. **The Cocker family who are in their 25th year** have developed an outstanding country house hotel with a string of awards to its name. You'll find a high standard of cuisine, complemented by well-chosen wines and friendly efficient service. Residents may prefer aperitifs in the secluded Resident's Lounge Bar, or a real ale in the Public Bar with its cosy log fire.

All 26, superbly appointed, en-suite guest rooms are equipped with every amenity, and four poster beds or balconies are available. Others, like those on the Museum Wing's ground floor, have easy access ideal for elderly guests, and families are well catered for with free beds for children.

Self-catering apartments - This new stone development adds suites, bedrooms and self catering apartments for up to 20 people. They have splendid open views over the bridge, the river and out to Dunterly Fell (The Pennine Way). Residents have free use

of all the Hotel's facilities, including the large, heated, indoor swimming pool and sauna. The surrounding countryside provides excellent walking opportunities, from the Pennine Way and Hadrian's Wall to Kielder Forest. Bellingham Golf course, opposite, hosts county matches and the hotel receives favourable terms. Being the nearest hotel to Kielder Water it is an ideal base for all yachtsmen and water sports.

Riverdale is the only hotel in the North of England with the prestigious 'Les Routiers' Gold Plate award

Orchard Guest House

High Street, Rothbury, Northumberland NE65 7TL Tel: 01669 620684
email: jpickard@orchardguesthouse.co.uk www.orchardguesthouse.co.uk

This charming country Guest House is over 200 years old, and although it has every modern and up to date convenience, it still retains the peace and character of a bygone era. The house overlooks the village green in Rothbury, a picturesque and thriving little community set in the heart of the beautiful Coquet Valley. With the rolling unspoilt countryside all around, and the heritage coastline just a few miles away, a memorable stay is most certainly yours, if you spend time with us! We have six bedrooms to offer our guests, all of which have washbasins, shaver points, colour television and tea and coffee facilities, most rooms have shower and toilet en suite. We are happy to provide sleeping arrangements to suit your own particular requirements - twin, double or family rooms. The house is fully centrally heated, and there is a comfortable lounge available during the day should the weather be unkind. It has a large and varied selection of books, as well as local information which we try to keep right up to date. In our comfortable dining room, you will be offered a full, freshly cooked English Breakfast to start your day. Orchard House is a non-smoking guest house.

Northumbria Coast & Countryside

The Northumbria Coast

To many visitors, the Northumberland coast is its most attractive feature, especially the long, un-crowded sandy beaches. There are also many attractive fishing villages and little seaside resorts to visit. The 40-mile stretch of coast from Amble to Berwick-upon-Tweed has been officially recognised as both a "Heritage Coast" and an "Area of Outstanding Natural Beauty" - to be conserved on behalf of the nation.

Northumberland has one of the cleanest coastlines in the country. In 2001, its 12 main beaches all passed the Environment Agency's basic standard for Coastal Bathing Water. Six beaches went on to achieve Rural Seaside Awards for their cleanliness and safety from the Tidy Britain Group.

Spittal Beach: Soft golden sands on the south side of the Tweed estuary at Berwick.

Bamburgh Beach: Mile upon mile of long, golden sands overlooked by the magnificent Bamburgh Castle.

St Aiden's Beach, Seahouses: An arc shaped sandy beach with rocky outcrops and rock pools offering good views of the Farne Islands.

Beadnell Bay: A golden, sandy horseshoe-shaped beach overlooked by sand dunes and a small harbour.

Embleton Bay: Reached from the villages of Low Newton and Embleton, this quiet, sandy bay is overlooked by the dramatic ruins of Dunstanburgh Castle.

Alnmouth: Soft, golden sands on the north side of the pretty, red-roofed holiday resort of Alnmouth.

Warkworth Beach: Miles of soft, golden sands backed by an extensive range of dunes, a short drive from the historic village of Warkworth.

Amble Links: An unspoilt beach with occasional rocky pools, sheltered by an extensive range of dunes.

Druridge Bay: A long, gently curving sandy beach.

Newbiggin Beach: The beach is framed by the town itself and its new sea walls and esplanade, overlooked by St Bartholomew's Church above a rocky outcrop.

Blyth, South Beach: Soft, golden sands to the south of Blyth Harbour, overlooked by a line of windmills on the harbour wall.

Seaton Sluice Beach: Long, golden sands stretching north from the picturesque harbour of Seaton Sluice.

There are many attractions along the Northumbria Heritage Coast and only lack of space prevents them all from being included, however you may consider a visit to:

Amble Marina. Northumberland's only

Marina for Yachts is at Amble. It is in a lovely location, to the west of the fishing harbour, bordered by the open grassland of Amble Braid, and with good views along the Coquet River to Warkworth Castle. The marina basin provides deep water at all states of the tide but a sandbar at the harbour mouth prevents access to

the sea at very low tides. There are a small number of available berths for both local and visiting craft.

Warkworth Castle, Standing on a hill and dominating the village of Warkworth, the dramatic ruins of the castle provide an evocative image of medieval strength. Norman in origin, the castle was taken over by the Percy family in 1332. Later that century, Henry Percy, the first Earl of Northumberland, allowed the castle to be the home of his eldest son, Harry Hotspur. The opening scene of Shakespeare's play "Henry IV" is set here. Alnmouth: A popular but peaceful coastal resort with superb sandy beaches and two golf courses, including the second oldest in England. The red roofed houses which flank the estuary of the River Aln are particularly picturesque.

Seahouses, a colourful and lively seaside resort and fishing harbour where you can take boat trips to see the bird and seal colonies on the Farne Islands throughout the summer. A Marine Life Centre is another attraction of the town and there are a wide range of hotels, guest houses and caravan and camping sites in the area.

The Northumbria Countryside

Move inland from the coast and the scenery is no less stunning. From the Durham Dales to the south, up through the Cheviot hills to the Scottish border in the north there is a never ending pageant of vistas, each competing for the attention of your eye like entrants in a beauty parade.

In the western march is Kielder Forest Park, the largest man-made forest in Europe surrounding Kielder Water. With a 27 mile shoreline and covering 2684 acres, this feat of engineering was created in 1976 by damming the North Tyne. There is a network of footpaths throughout the forest and around the reservoir and there are centres for visitors, field studies, fishing and sailing. Wildlife have a relatively safe haven in this part of England, it is one of the few places that red squirrels can still be found. Bolam Lake, created by John Dobson in 1818, is now surrounded by a country park that is home not only to the red squirrel but many native and migratory birds throughout the year.

Alston, said to be England's highest Market town at an altitude of 1,000 feet above

sea level, looks out over the counties of Durham, Northumberland and Cumbria. Weathered stone houses cluster around the narrow cobbled streets as if to protect the community from the harsh winter winds.

You can get a feel for what it was like to live in Weardale during the late 1800s. The rugged past of the upper Wear Valley is shown in the Weardale Museum at the High House Chapel, near Ireshopeburn on the A689, built in 1760 this is the second oldest Methodist chapel still in use. Northumbria is such a vast area of natural beauty it's a shame that there is so little space in which to describe it, every twist and turn of the country lanes reveals many of the areas beautiful secrets, from cottages to cathedrals, castles to chapels a trip through Northumbria will never dissappoint. So why not get out there and see for yourself!

Mizen Head HOTEL

Bamburgh
Northumberland
NE69 7BS
Tel:01668 214254
Fax: 01668 214104
e-mail: reception@themizenheadhotel.co.uk
www.themizenheadhotel.co.uk

Welcome to the beautiful coast of Northumberland. Rugged cliffs and swathes of sandy beaches skirt the high moors and hillsides of these borderlands. The pretty 18th century cottages of Bamburgh are tucked in safely beneath the multi turreted walls of the 12th century castle. The huge curtain wall and keep are perched defiantly on top of a 150 foot coastal crag. From the battlements you can enjoy an uninterrupted view of the Cheviot hills to the west and the Farne islands to the east. The delightful village also boasts a museum devoted to the memory and heroic feats of Grace Darling and her father, the lighthouse keeper, who rowed out together to rescue shipwreck survivors in 1838.

The Mizen Head Hotel provides the perfect base from which to explore this beautiful part of the country. At this family-run hotel all the guests are treated as old friends by the staff. Always happy to help, nothing is too much trouble, wether its planning a trip, looking for local information or ensuring that guests particular dietary needs are catered for.

The family theme runs throughout the hotel as there is a well equipped playground and a large beer garden adjacent to the conservatory for those pleasant family days out. Family rooms are available with a baby listening service and cots if required. Pets can also be

available. Relaxation is the main objective at the Mizen Head, the location is perfect for walkers; painters; photographers and, for the more energetic there are some superb golf courses all within a few minutes drive, booking your round or party through the hotel will get you an additional discount at the course.

At the end of an enjoyable day the bar is just the place to raise a glass and

accommodated by prior arrangement. The fresh air is like wine and certainly sharpens the appetite for a meal in the Mizen Head's excellent restaurant. The menu reflects the seasons and the chef only uses fresh local produce to produce a selection of dishes both traditional and exotic. If its a quick snack in the bar, traditional Sunday lunch or dinner from the a la carte menu in the restaurant you will not be disappointed, there is always a children's menu and vegetarian options

contemplate the evenings entertainment. The hotel is a popular venue for local, talented musicians and many great evenings have been enjoyed by the guests - but don't worry they won't disturb your nights sleep in one of the comfortable rooms!

Finding the Mizen Head couldn't be easier from either north or south, just turn onto the Bamburgh road at the Purdy Lodge services on the A1.

Longstone House Hotel

182 Main Street - North Sunderland - Seahouses - Northumberland - NE68 7UA
Tel: 01665 720212 Fax: 01665 720211 e-mail: info@longstonehousehotel.co.uk
www.longstonehousehotel.co.uk

The Longstone House Hotel is situated in the quiet, old part of Seahouses, which is called North Sunderland. We are only a few minutes away from miles of Heritage Coastline which carries the Blue Flag Award for clean beaches. Nearby is the picturesque fishing harbour of Seahouses, from where one can take boat trips to visit the bird sanctuaries and seal colonies of the Farne Islands.

The hotel is English Tourist Board '2 star' standard commended. We have two bars, the Bamburgh Bar with a pool room, darts etc., and in the comfortable Lindisfarne Lounge bar there is a log burner on cold days and evenings.

Our local chef David Barella has presided over the kitchens of Longstone House for the past few years. He produces superb meals that can stand comparison to any that may be found in a large city.

We have eighteen bedrooms which are all en suite with either a bath and/or shower. Rooms are comfortable with central heating, tea and coffee facilities and colour televisions. Many rooms are double glazed. We have double, single, twin bedded and family rooms and also have one 2 bedroomed Family Suite.

The area has a wide range of activities for the outdoor enthusiast with superb beaches, birdwatching, horse-riding (we can arrange special riding breaks and limited stabling is available if booked in advance), sailing, diving, hiking, golfing (there are 8 golf courses within 25 miles of the hotel, special rates are available for many if booked through us), tennis, sea & river fishing. Please do not hesitate to contact us if we can assist in planning your holiday in Northumberland.

The PACK HORSE INN

Ellingham, Chathill, Northumberland Tel: 01665 589292
e-mail: thepackhorseinn@hotmail.com

Less than a mile from the dual-carriageway section of the A1 at Brownieside, lies the picturesque village of Ellingham. The Pack Horse stands at one end of the village street and has graced the village for over 180 years, it provides a congenial meeting place for villagers and visitors alike.

After a busy day exploring the many sites of this exceptional county, the bar, dining room and pool room of the Pack Horse offer welcome relaxation for the whole family, where you can mingle with the locals.

It provides real ales, excellent beer and spirits which may well be helped down by the home-cooked fresh food and seasonal vegetables provided in the tasty bar lunches and dinners. Vegetarians receive equal consideration and special requirements are catered for, or look out for the Chef's specials on the blackboard. All meals are served in our cosy dining room.

The Pack Horse provides superior Bed and Breakfast accommodation and can now offer five en-suite bedrooms. Two are double and three twin-bedded, four have showers and one a bath. All rooms are fully equipped with colour television and have hot drinks trays, and the clock alarm radio will wake you in plenty of time to enjoy your full English breakfast, cooked to your requirements.

If you or your party would like complete privacy, we have the adjoining cottage available to let on a weekly basis, Saturday to Saturday. Please ask about the rates. it will accommodate two to four people, and has one bedroom. The bed-settee in the sitting room provides the other two if needed. The cottage has full night storage heating, electric cooker, microwave and fridge-freezer. The bathroom has a bath with an electric power shower. There is ample parking. The Pack Horse is situated 42 miles north of Newcastle and 77 miles south of Edinburgh.

The Northumberland Estates Holiday Cottages

Alnwick Castle - Alnwick - Northumberland - NE66 1NQ

Holiday Heaven!

On the Duke of Northumberland's Estate we have peaceful stone cottages. Enjoy the large sandy beaches at Bamburgh and the Northumberland National Park. With intriguing castles to explore, islands to visit, hills to roam and rock pools to investigate it is an ideal area for all the family to enjoy a relaxing holiday. From a cottage for 2 opening onto the beach, or an Old Mill offering luxury for up to 12.
All 17 cottages are ETB 4 or 5 star and are open all year. Each cottage is superbly equipped, including dishwasher, telephone, video, four poster bed and en suite facilities. Dogs by arrangement. 3 night winter breaks available. Close by is our own small country pub the Apple Inn offering excellent food and a friendly welcome.

Jane Mallen
Alnwick Castle
Northumberland
NE66 1NQ
Telephone 01665 602094
Fax 01665 608126
Email nehc@Farmline.com
Website alnwickcastle.com/holidaycottages

44 Northumberland Street
Northumberland NE66 2RA

Alnmouth
Tel: 01665 830363

e-mail: debbiephilipson@hopeandanchorholidays.fsnet.co.uk
www.hopeandanchorholidays.co.uk

The Hope & Anchor is reputed to be one of the oldest buildings in Alnmouth village and although originally two cottages, it has been an Inn for as long as the present villagers can remember and its walls will not doubt carry secrets and stories, which only they can tell.

Today it is under the ownership of George and Debbie Philipson, whose main aim is to provide a comfortable base from which the guest can enjoy the delights of the heritage coastline, golf courses, cycle ways, walks and wide sandy beaches. Within a few miles radius of the hotel are Alnwick Castle (where the 'Harry Potter and the Philosophers Stone' movie was filmed) and the newly opened 'Alnwick Garden' with its fascinating water features. Warkworth, Dunstanburgh and Bamburgh castles are all a short drive away and being on an

excellent bus route and close to the Alnmouth mainline railway station, makes Edinburgh & Newcastle a good day out too. The location is perfect for the car free traveller. George and Debbie provide a special welcome to all of their guests, with hearty breakfasts, picnic lunches (by prior arrangement) and a cosy well stocked bar. The atmosphere is relaxed and without undue formality. A large cottage is also available, it sleeps 8 in comfort and is just a short walk from the Alnwick Garden.

The staff are well versed in their knowledge of the surrounding area and can help you to plan your stay, booking boat trips, horse riding, checking out the opening times of the historic houses and castles and safe crossing/tide times for Holy Island - anything to make your visit as pleasant as possible.

Haggerston Castle Holiday Park, Near Berwick Upon Tweed, Northumberland, TDI5 2PA.
Telephone: 01289 381333.

Lovely landscaped parklands sweep down to extensive lakes & a picture book tower....Haggerston Castle was once a stately home! With Northumberlands' National Park, wonderful beaches, historic castles & so much to do on park as well, you will be spoilt for choice every day! With sensational entertainment for all the family it's no wonder Haggerston Castle has been chosen by two TV holiday programmes.

BRITAIN'S LEADING INDEPENDENT HOLIDAY PARK COMPANY

What's on Park......

* Indoor & Outdoor Heated Swimming Pools
* Family Entertainment & Kids Clubs
* Evening Shows & Cabaret
* Amusement Arcade
* Pool & Snooker*
* Mini Ten Pin Bowling*
* Tennis Courts & Bowling Green
* Par 3 9 Hole Golf Course*
* Boat Hire*
* Horse Riding*
* Bike Hire*
* Children's Outdoor Play Area

No need to cook!....

* Aqua Bar for all day meals & snacks
* Foodworks Fish & Chip Shop
* Castle Café for snacks
* Burger King
* Chinese Restaurant
* Chinese Takeaway
* Supermarket with bakery & off licence

Special to the Park....

Haggerston Castle was chosen by both 'Wish You Were Here' & the BBC 'Holiday' programme....& you'll love it too...stroll around our landscaped parklands, feed the wildfowl & ducks on the lakes & take the boat out on the lake.

Spring & Autumn breaks from only £18.00 per person (4 share)
Summer weeks from only £67.00 per person (6 share)

For more information & a free colour brochure call now on 01289 381333.

* Subject to an extra charge

CHARLIES

Albert Street, Amble, Northumberland
Telephone - 01665 710206

Nestling in the attractive Northumberland fishing town of Amble, close to the historic Warkworth Castle, you will find Charlie's Fish and Chip Take-Away and Restaurant. In 1986 Joyce and Charlie Willoughby started the business having been dissatisfied with their careers in teaching and taxi driving. They started as complete novices but with a little bit of the Midas touch plus a lot of passion and enthusiasm they have built a business to be proud of. People now travel far and wide on a regular basis to enjoy the finest of cuisine.

The extensive menu not only caters for fish and chip lovers. There is a diverse range including vegetarian and healthy eating options and children have their own menus in both the take-away and the restaurant.

The restaurant seats 24 but you can always take away and sit at the nearby harbour or marina.

The story doesn't end there. Due to the enthusiasm, training and teamwork of Joyce, Charlie and their staff they have won many business and catering awards. Awards and training certificates line the wall of the shop.

To learn more about "Charlies" log on to their web site at http://www.charlieschips.com. There are pages on the background of the company, its location, what can be seen in the area, menus and "meet the staff"

Frying Times
Sunday to Wednesday - 4.30pm to 9.00pm
Thursday, Friday and Saturday -11.30am to 9.00pm
Bank Holiday Mondays - 11.30am to 9.00pm

INVESTOR IN PEOPLE

DOXFORD FARM COTTAGES COUNTRY STORE & COFFEE SHOP

Chathill - Alnwick - Northumberland - NE67 5DY
Tel: 01665 579348 Fax: 01665 579331 e-mail: doxfordfarm@hotmail.com
www.doxfordfarmcottages.com

Doxford farm is a 500 acre working farm situated amidst undulating, unspoilt wooded countryside. The lake and its small ponds hold brown trout, the farm also has five miles of wooded and lakeside walks and trails.

Doxford Farm Cottages provide the perfect base for a relaxing holiday in the stunningly beautiful county of Northumberland. There are 7 stone built cottages, each with its own unique character. They vary in size from one to three bedrooms with a mixture of double, twin and family rooms. All have excellent facilities such as fridge, washing machines, microwave etc. plus tv and video. Some have traditional open fires and some with wood burning stoves. No matter how big or small your party, you are guaranteed a comfortable and enjoyable stay.

The Doxford Country Store and Coffee Shop is part of the complex and provides a huge range of gifts from all over the world including cotton throws and rugs, recycled glassware, rustic baskets, candles, teddies and country American paints. We also have lots of Cottage Delight food products. The coffee shop has a wide range of cakes, sandwiches, teas, coffees and soups. You can have the choice of eating indoors or outdoors in our beautiful courtyard.

HECKLEY HIGH HOUSE
FARM COTTAGES

Tel: 01665 602505 Fax: 01665 606945 - E-Mail: rlgreen@heckley.fsbusiness.co.uk

Richard and Susan Green and family assure you of a warm welcome to their mixed family farm in Northumberland.

Situated about one mile north of the historic market town of Alnwick your cottage location is an ideal centre from which to explore the many attractions and unspoilt countryside the district has to offer.

The magnificent Alnwick Castle, with its Harry Potter film connections, and the new Alnwick Garden are only a few minutes drive away. Explore our beautiful coastline with its many quiet sandy beaches, visit castles at Bamburgh, Holy Island, Warkworth, Dunstanburgh and Chillingham. Venture inland and discover wonderful stately homes such as Cragside and Wallington or walk in the Cheviot Hills. We have three cosy and comfortable well

equipped cottages to offer. They are all south facing with splendid views over open countryside and have private enclosed lawned gardens with garden furniture. They have log fires and all fuel and electricity are included as are bed linen and towels. (Cot and high chair available on request)

We are happy to welcome well behaved pets and horse livery is also offered.

Cheviot View Sleeps 6 (1 double, 1 twin, 1 bunk room) Oil fired central heating, washing machine, tumble dryer, microwave, TV & Video, telephone and bathroom with electric shower over bath.

Paddock Cottage Sleeps 5 (1 double, 1 twin, 1 single) Oil fired central heating, dishwasher, washing machine, tumble dryer, microwave, TV & Video, telephone and bathroom with electric shower over bath.

Dipper Cottage Sleeps 4 (1 double, 1 twin) Night storage central heating, washing machine, tumble dryer, microwave, TV & Video, telephone and bathroom with electric shower over bath.

Guests are welcomed on arrival to their cottage with one of Susan's chocolate cakes or special chocolate puddings, which are home made at the farmhouse bakery.

Please contact Richard and Susan for details and availability.

i TouristInformationBritain.com *i*

Your first stop on the web for information on:
WHAT TO SEE! WHERE TO GO! WHAT TO DO!
IN GREAT BRITAIN

Shop On-Line

Check out our range of excellent quality items

www.TouristInformationBritain.com

ISBN 1-902708-01-6

Acknowledgments - Copyright - Disclaimers

Circular Walks in Derwentside

There are many walks and points of interest located within the district. Spectacular landscape and memorable scenery are hallmarks of the area, which is home to some of the most beautiful and least spoilt countryside in Britain. Our circular walks take you through some of the most beautiful parts of the North East, so why not plan a short break, pull on your walking boots, try one of the six routes and discover the hidden beauty that is Derwentside.

Route 1 Allensford - Shotley Bridge - Allensford: Offers changing scenery, taking the walker through peaceful woodland to finish with spectacular views over the Derwent Valley.
Time: 1 1/2 - 3 Hours
Distance: 3 1/2 - 5 1/2 Miles

Route 2 Allensford - Castleside - Allensford: Set in the beautiful Derwent Valley taking in riverside woodland to finish with a magnificent view back to Allensford Country Park.
Time: 1 1/2 - 2 Hours
Distance: 2 1/2 - 4 Miles

Route 3 Allensford - Hownsgill Viaduct and Lydgetts Junction - Allensford: Takes the walker through the landscaped old Consett Steelworks Site The downhill stretch gives commanding views over the Derwent Valley.
Time: 1 1/2 - 2 Hours
Distance: 4 1/2 miles

Route 4 Derwentcote - Medomsley - Derwentcote: Set in the beautiful Derwent Valley, taking the walker uphill through woodland and fields to experience panoramic views at its peak.
Time: 2 1/2 - 3 Hours
Distance: 4 1/2 - 6 1/2 Miles

Route 5 Burnopfield - Low Friarside and Lintz Green - Burnopfield: Leafy country lanes provide access uphill to emerge above Burnopfield village where a downhill sweep of picturesque woodland awaits.
Time: 2 1/2 - 3 Hours
Distance: 6 1/2 Miles

Route 6 Moorside - Knitsley - Rowley - Moorside (incorporating 3 shorter routes): Starting in the shadow of Hownsgill Viaduct and giving the walker the option of taking one of three walks.
Time: 1 1/2 - 3 Hours (depending on route chosen)
Distance: 3 - 6 Miles (depending on route chosen)
For further information, please contact:
Project Development Officer for Tourism
Economic Development and Regeneration Unit
Derwentside District Council
Civic Centre
Consett
Co. Durham
DH8 5JA
Tel: (01207) 218237
E-mail: info@virtualtourismcentre.com
Web Site: www.virtualtourismcentre.com

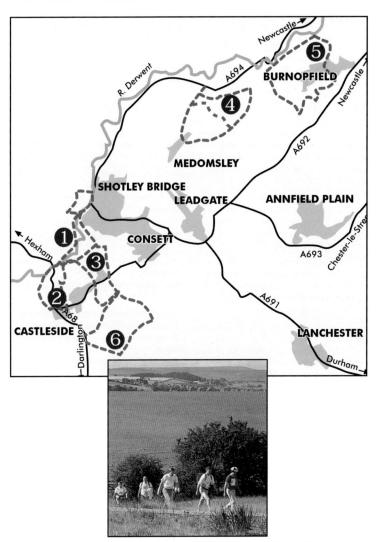

Reproduced from Ordnance Survey Mapping on behalf of The Controller of Her Majesty's Stationery Office © Crown Copyright licence No 100026178

Discover Derwentside by Bicycle

Whether you are a serious cyclist or a family group, one of the best ways to explore Derwentside is by bicycle. At present there are seven dedicated cycle paths within the district, passing through both urban and rural areas. These routes largely follow the former railway lines, which have been reclaimed to provide an excellent network of accessible and traffic free routes for cyclists, walkers and horse riders. Four of these routes form part of the National Cycle Network which now extends across much of the UK. These routes intersect at Lydgetts junction west of Consett, which is an excellent access point with a free car park and picnic area. Other car parks with picnic sites are shown on the map opposite. The Derwent Walk, Waskerly Way and the Consett and Sunderland Railway Path form part of the award-winning Sea to Sea (C2C) Route, a 140 mile trail linking the west and east coasts of England which offer a challenging ride for the more experienced and adventurous cyclist.

Derwent Walk Cycle Path - Length 14 km (9 Miles) - NCN route 14

From Lydgetts Junction, this main path runs downhill through the picturesque Derwent Valley to the River at Swalwell via Rowlands Gill. Between Shotley Bridge and the Tyne it runs through a country park, which includes the most scenic parts of the route where it passes Ebchester, with its Roman Fort, and crosses the spectacular Pont Burn Viaduct. The National Trust's Gibside property lies a short ride off the route on the road between Burnopfield and Rowlands Gill.

Instead of crossing the reclaimed Consett Steelworks site at Berry Edge, riders have the option of using a new cycle path running through Consett Town Centre (if you need refreshments or shopping) and Consett Park. Look out for the huge 'hot metal' carriages, once used to carry 50 ton loads of molten iron, which are stationed along this new path, and the awe inspiring, stainless steel 'Terris Novalis' sculptures. For further information on the following cycle routes please contact Derwentside District Council's Project Development Officer for Tourism (details below) for a free leaflet:

Waskerley Way - Length 15 km (9 Miles) - NCN route 7

Lanchester Valley Walk - Length 14 km (9 Miles) - NCN route 14

Consett and Sunderland Railway Path - Length 16 km (10 Miles) - NCN route 7

Bowes Railway Path - Length 4 km (3 Miles)

South Stanley Green Corridor - Length 4 km (3 Miles)

Kyo Greenway - Length 5 km (3 miles)

Contact Details:

Project Development Officer for Tourism
Economic Development and Regeneration Unit
Derwentside District Council
Civic Centre
Consett
Co. Durham
DH8 5JA
Tel: (01207) 218237
E-mail: info@virtualtourismcentre.com
Web Site: www.virtualtourismcentre.com

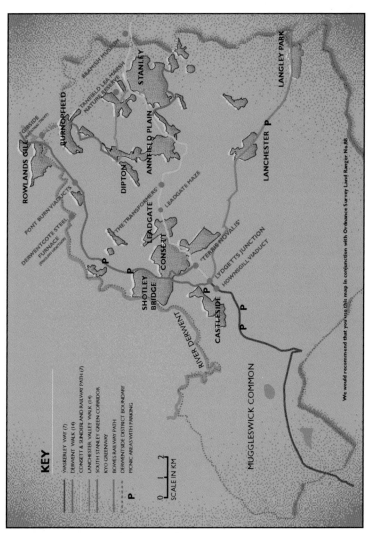

KEY

WASKERLEY WAY (7)
DERWENT WALK (14)
CONSETT & SUNDERLAND RAILWAY PATH (7)
LANCHESTER VALLEY WALK (14)
SOUTH STANLEY GREEN CORRIDOR
KYO GREENWAY
BOWES RAILWAY PATH
DERWENTSIDE DISTRICT BOUNDARY
P   PICNIC AREAS WITH PARKING

0   1   2
SCALE IN KM

We would recommend that you use this map in conjunction with Ordnance Survey Land Ranger No.88

LANGLEY PARK
LANCHESTER P
BEAMISH MUSEUM
STANLEY
TANFIELD LEA MARSH NATURE RESERVE
BURNOPFIELD
ANNFIELD PLAIN
DIPTON
ROWLANDS GILL
GIBSIDE (National Trust)
PONT BURN VIADUCTS
DERWENTCOTE STEEL FURNACE (English Heritage)
THE TRANSFORMERS
LEADGATE
LEADGATE MAZE
CONSETT
'TERRIS NOVALIS'
LYDGETTS JUNCTION
HOWNSGILL VIADUCT
SHOTLEY BRIDGE
RIVER DERWENT
CASTLESIDE
MUGGLESWICK COMMON

Drive Through the Land of Prince Bishops

The route described here can be joined at any point, it is around 100 miles long and provides a full days outing.

Point 1. Durham City.
Home of Durham Cathedral and Castle. The Castle became the foundation college of the University in 1836.

Point 2. Lanchester.
A pleasant village on the line of the Roman Road. Dere street (Motor Trail leaflet available).

Point 3. Wolsingham.
The gateway to Weardale. Explore this attractive Dales market town on foot.

Point 4. Stanhope.
Capital of Weardale. Look out for the fossil tree stump in the churchyard, Visit the Dales centre.

Point 5. Westgate and Eastgate.
Once marked the boundaries of a park about 13 miles around, set aside by the Bishops in the early 13th century for the hunting of deer.

Point 6. Killhope Lead Mining Centre.
Years of painstaking restoration has made this the best preserved lead mining site in Britain. It stands as a monument to the lead mining families.

Point 7. High Force.
The highest waterfall in England.

Point 8. Bowlees.
A picnic area complete with its own waterfall.

Point 9. Middleton - In - Teesdale.
The Capital of Upper Teesdale and once the northern HQ of the Quaker London Lead Company. The cast iron canopied drinking fountain is a tribute to one of the company managers.

Point 10. Barnard Castle.
Roman river crossing point. Norman stronghold. Visitors to the town include Richard III, Cromwell and Dickens. Home of the Josephine and John Bowes Museum.

Point 11. Raby Castle.
One of the grandest castles in England. A Neville stronghold for 400 years, it passed to the Vane family in 1626.

Point 12. Bishop Auckland.
A market town which grew around the Bishops favourite country residence. The Castle, with its 800 acre park, is now the official residence of present day Bishops.

For more detailed information collect the Drive through County Durham leaflet from any Tourist Information Centre within the County.

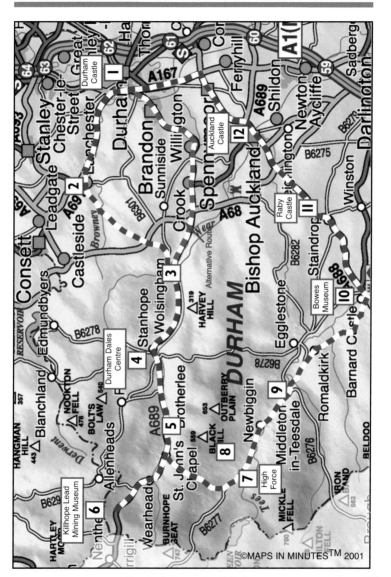

Shincliffe Wood and Riverside

The walk is 5.6km (3¹/₂ miles)

Start at The Post Office, High Shincliffe. The adjacent houses are all that remain of the original settlement. The colliery village has long gone and has been replaced by modern housing.

Walk away from the A177 and take the first left turn signposted to Shincliffe Primary School. Walk along the road until you come to the open fields on your left. Turn left at the footpath, through farmland, signposted to Manor House Farm. Continue a short distance and, immediately after passing a hedge, turn left and follow the footpath down towards Shincliffe Village. At the bottom of the hill rejoin the A177 and turn right.

Keep on the main road until you reach the church. Follow the footpath through the church grounds which takes you to the centre of the village. Shincliffe is a typical Durham 'green' village. It is particularly picturesque during the spring when masses of daffodils bloom on the village greens. Turn right and walk to the end of the village then turn left at the sign post to the garden centre which was once the site of Shincliffe Mill.

Continue along the road until Shincliffe Hall is reached. It is used as University accommodation and was built in 1781 by William Rudd, recorder of Durham. Local materials of handmade bricks and a stone slab roof have been used. Just before Shincliffe Hall there is a footpath to the left which is signposted to "High Butterby Farm". Follow this path through Shincliffe Wood. The mixed woodland provides a good habitat for a wide variety of flora and fauna. Continue along the footpath through the wood beside the riverbank climbing the steps up the hillside before returning to the riverbank. Continue across a bridge and climb a steep hill alongside a fence. At the top of the hill the path comes out of the wood. This is the track for High Butterby Farm.

On reaching the tarmac lane turn left and continue. At the crossing of the paths turn left. This ancient bridleway is called Strawberry Lane. Still used as a footpath between Ferryhill and Durham, it was once trodden by drovers, peasants and possibly Roman legions.

On your left is the grandstand of Shincliffe Racecourse which has long been used to store hay but some of the tiered stands remain. The concrete block supported the starter's bell. The course was built to replace Durham City Racecourse which had become University playing fields. The venture failed because of the onset of the 1914 -18 war.

Before High Shincliffe comes back into view you will see High Grange Farm with its red pantiled roof. Now extensively renovated, the house is thought to be the village's oldest building. The narrowness of the house indicates its extreme age because early builders had difficulty in roofing wide buildings. The derelict building to the left housed a powder magazine where explosives for Bowburn Colliery were stored.

A tarmac road returns to the A177 at High Shincliffe. Cross the road with care to return to the starting point.

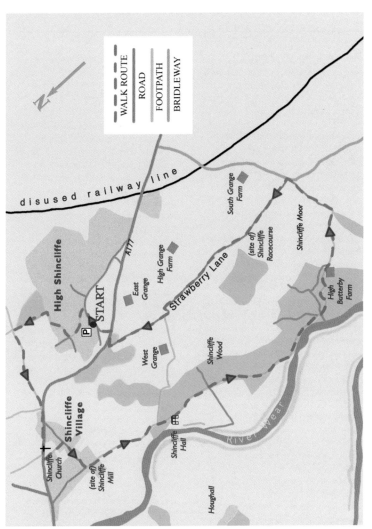

Reproduced from Ordnance Survey Mapping on behalf of The Controller of Her Majesty's Stationery Office © Crown Copyright licence No 100026178

Low Force & Gibson's Cave

The walk is 4km (2¹/₂ miles)

Start at the Bowlees picnic area car park. From the car park pass behind the information hut, go over the cattle grid and along a track leading to Summary Hill Farm. Follow this for a short way uphill and, where the track bends left, leave it and head right towards a gate.

After Hood Gill cross a stone bridge, pass to the right of the barn and climb to see the view at a stile. There are larch and pine plantations; mature ash and sycamore with a few oak, and beyond, the ancient High Force juniper forest.

Climb over a stile next to a field gate and follow the road down to Newbiggin, once a lead smelting village when the North Pennines was the most productive lead mining area in the world.

Turn right down the narrow road through Fellowship Farm. *The road is narrow - keep to single file when traffic is passing.* At the main road turn right and walk a short distance to a stile on the opposite side. Grass is grown in these meadows for summer haymaking to feed sheep and cattle throughout the long hard winters. The hay is stored in barns called 'byres' or 'laithes', like the one over the wall to your right.

Cross the Bow Lee Beck. A great bank of beech trees are growing on a limestone ridge to your right. Pass through a field gate by an old barn. On the skyline to your left is the tree-crowned point of Kirkcarrion, the site of a bronze-age burial mound. Crossing Scoberry Bridge look for the 'circular' kettles in the flat rock below.

They are caused by hard pebbles being swirled round and round by swift water so that the softer shales are ground away. Turn right onto the Pennine Way, which you will follow for about 800 metres (¹/₂ mile) along the course of the Tees. As you walk upstream you can see glimpses of Holwick Lodge. Cross a wooden stile; to your left there are mediaeval ironstone mines. Cross a wooden bridge and look for Staple Crag, a great whinstone buttress. From here the River Tees narrows to a rocky gorge under Wynch Bridge at the foot of Low Force. 'Force' is a corruption of the old Norse word 'foss' for waterfall. A local name for Low Force is "Salmon Leap". The Tees was a salmon river before it became polluted in its lower reaches.

Follow the path from the river to a stone stile at the edge of the wood. These massive whinstone buttresses are known locally as 'Fat Ladies Torment'. The path goes through the meadow to a kissing gate at the road. Cross carefully and go up the narrow lane to Bowlees. Most buildings are whitewashed, indicating Raby Estate ownership.

The only building entirely in plain stone is the old chapel, now Bowlees Visitor Centre run by Durham Wildlife Trust. If you wish to extend the walk by ¹/₂ a mile follow the signs to Gibson's Cave. The cave is behind Summerhill Force. Gibson was a 16th century outlaw who hid and lived behind the waterfall when pursued by constables of Barnard Castle.

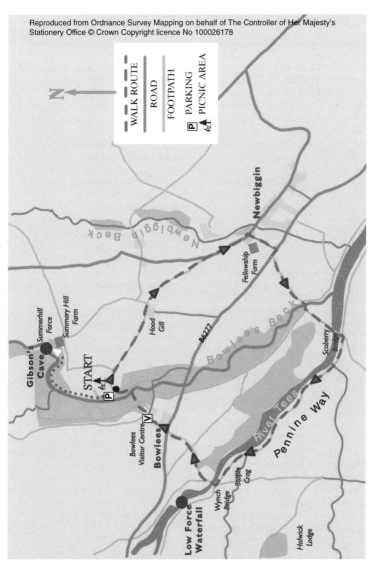

WALK ROUTE
ROAD
FOOTPATH
P PARKING
PICNIC AREA

N

Newbiggin

Newbiggin Beck

Summerhill Force

Gibson's Cave

Summerhill Farm

START

Fellowship Farm

Hood Gill

B6277

Bowlees Beck

P

Bowlees Visitor Centre

Bowlees

Scoberry Bridge

River Tees

Pennine Way

Wynch Bridge

Scoberry Crag

Holwick Lodge

Low Force Waterfall

Coastal Walk - South Tyneside

**CYCLING AND WALKING
IN SOUTH TYNESIDE
COASTAL ROUTE (7 miles)**

Section 1 From the South Groyne walk along Little haven Beach to the start of the pier where you need to turn left for approximately 200 metres before turning right along the promenade. Continue past the fairground and the amphitheatre until you reach the far end of the bay at Trow Point. Take the stone track on your left signed 'Coast Footpath'. Follow the stone path next to the cliffs over the Leas area past the former Trow Quarry and Frenchmans Bay.

Points of Interest
* The Conversation Piece
Created by the late Spanish artist Juan Munoz in 1999 the 22 bronze life size figures command a superb view of South Shields harbour.

* The Leas
Managed by the National Trust - 300 acres of grassy open space is home to ground-nesting birds and wild flowers.

Section 2 When you reach the Northern end of Marsden Sands you have two options:

1) Follow the coastal path along the top of the cliffs to Marsden Rock.

2) Walkers could go down the slope onto the beach and back up the 100 steps near the Tavistock at the Grotto Bar, Bistro and Restaurant. A lift provides walkers with access to the cliff top.

From Marsden Rock follow the grass coastal path towards Souter Lighthouse past the Lime Kilns. When you reach

Souter Point car park access road, turn left towards car park and follow the stone path heading south by the side of the lighthouse. Go past Souter Lighthouse and continue along the coastal path down to the Rifle Range.

Points of Interest
* Marsden Bay and Rock
Spectacular cliffs, caves and arches and one of England's most important sea bird colonies.

* Marsden Lime Kilns
Built in the 1870's, the limestone from the Marsden Quarries was burnt in these kilns before being transported by train to South Shields. The Kilns ceased production in 1960.

* Souter Lighthouse
The first lighthouse constructed to use electric light, built in 1871.

Section 3 Once at the rifle range it may be necessary to follow a diversion.

If the range is in use: Red flags will be flying and you will not be able to walk this section. Instead take the path along the rifle range and Wheathall Farm. After the rifle range, follow signs for the National Cycle Network (NCN) Route 1 through the housing estate and you will rejoin the coast path at Whitburn Bents.

If the range is not in use: Red flags will not be flying and you will be able to continue straight down the coast. After leaving the rifle range area, continue along the coastal path at the rear of Whitburn Village until you reach the South Bents car park.

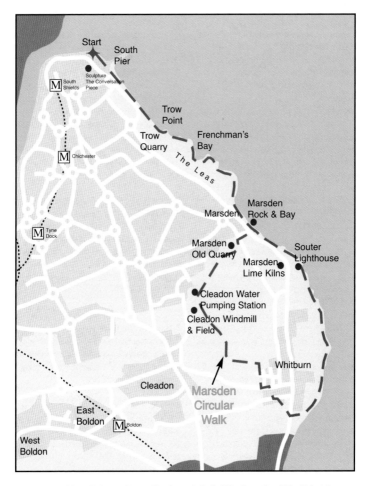

Start
South Pier
M South Shields
Sculpture The Conversation Piece
M Chichester
Trow Point
Trow Quarry
Frenchman's Bay
The Leas
M Tyne Dock
Marsden
Marsden Rock & Bay
Marsden Old Quarry
Souter Lighthouse
Marsden Lime Kilns
Cleadon Water Pumping Station
Cleadon Windmill & Field
Marsden Circular Walk
Whitburn
Cleadon
East Boldon
M Boldon
West Boldon

Reproduced from Ordnance Survey Mapping on behalf of The Controller of Her Majesty's Stationery Office © Crown Copyright licence No 100026178

Marsden Circular Walk - South Tyneside

A seven-mile route over Cleadon Hills and along the sea cliffs passing Windmill Field, Marsden Old Quarry and Marsden Rock. Follow the instructions detailed on the Coastal Route, section 2 and 3 to complete this route.

Section 1 From South Bents car park, cross the road and take the public footpath that runs parallel to the park up to Whitburn Village. The path brings you out onto Church Lane, with its 13th century parish church. The land emerges onto Front Street, cross the road and turn left before turning right up Sandy Chare and onto North Guards. On the left is the village pond and behind that is the former schoolhouse.

Bear right though and turn left up Wellands Lane. Follow this for a third of a mile and then take the public footpath on the left leading to Well House Farm. Follow the track past the farm and turn right along the way marked path that leads up to Cleadon Hills. From here on a clear day, you can see south to the Cleveland Hills and the coastline near Whitby.

Section 2 Continue along the path past 'Old Mans Garden' until you reach Cleadon Windmill, then follow the path next to the stone wall up to and through the kissing gate. On your left is Cleadon Water Pumping Station. Continue following the stonewall, until you reach another kissing gate, which gives you access to the golf course. Cross the course and go through the stone stile on the far side into Marsden Old Quarry. From here on a clear day, you can see north to the coast near Blyth and Cheviot Hills.

Point of Interest
* Cleadon Windmill and Field

Built in the 19th century, it was used for target practice by artillery until the First World War. Restored by Groundwork South Tyneside it is now a listed building. The field is of national importance for its wildflowers. Look out for beautiful plants such as the knapweed, salad burnet and autumn gentian.

Section 3 Turn right along the wall and follow the path that runs at the top of the old quarry towards the coast until you reach Lizard Lane. Cross the road and take the path between the Caravan Park and the golf course, which will bring you onto the Coast Road just south of Marsden Rock.

Point of Interest
* Marsden Old Quarry

A fascinating range of habitats including cliffs, woodland and grasslands. During the autumn the site is a magnet for migrating birds and bird watchers.

For information on cycling and walking in South Tyneside contact the Tourist Information Centre on 0191 4546612. For details on summer cycle hire telephone 0191 4247981.

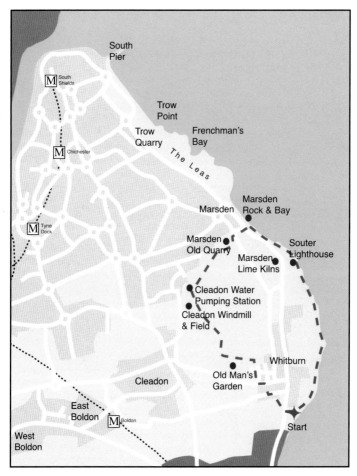

South
Pier

M South
Shields

Trow
Point

Trow
Quarry

Frenchman's
Bay

The Leas

M Chichester

Marsden
Rock & Bay

Marsden

M Tyne
Dock

Marsden
Old Quarry

Souter
Lighthouse

Marsden
Lime Kilns

Cleadon Water
Pumping Station

Cleadon Windmill
& Field

Whitburn

Old Man's
Garden

Cleadon

East
Boldon

M Boldon

Start

West
Boldon

Reproduced from Ordnance Survey Mapping on behalf of The Controller of Her Majesty's
Stationery Office © Crown Copyright licence No 100026178

Notes

Notes

CAPTURE THE SPIRIT OF

SOUTH TYNESIDE